Planting Guide to the Middle East

A. W. Van-Ollenbach

The Architectural Press: London
Nichols Publishing Company: New York

Mommersteeg International

First published 1978 by
The Architectural Press Limited.

Published in the United States 1978 by
Nichols Publishing Company, 155 West 72nd Street,
New York, NY 10023.

ISBN: 0 85139 513 9 (British Edition)

Designed & Printed by Parkway Illustrated Press, Abingdon, Oxfordshire.

Contents

Illustrations

Preface

Saudi Arabia and the Arab Gulf States cover a vast region and although a certain amount of data in connection with oil production, politics and early exploration is available, relatively little information with regard to environmental conditions and the establishment and maintenance of suitable planting material has been produced.

As this very large area has come into prominence during the last decade and enormous development programmes are envisaged, and are being undertaken, it is preferable from every point of view if as much information as possible is made available so that environmental design and planning can be facilitated within the limitations imposed by a hostile environment.

Generally speaking, this low rainfall region is hot and dry; brisk to high winds are experienced throughout the year, and it is often assumed that the range of vegetation is restricted and that there is a marked varietal similarity. Although this assumption is partially true, the vegetation is Saharo-Sindian in type and can be sub-divided as follows:-

(a) Vegetation existing under exceptionally harsh desert conditions
Areas such as the Rub al-Khali where precipitation is spasmodic and seldom if ever exceeds 50 mm, vegetation is limited to plants capable of surviving long periods of drought and most of the species that do exist under these conditions are very deep-rooted and produce seeds that can remain viable for many years.

(b) Vegetation existing under desert conditions
In areas where precipitation varies between 70 mm-100 mm, desert vegetation is relatively widespread and the range of plants is wider than it is in the Rub al-Khali. A large proportion of the vegetation that is found in central and eastern Arabia, the Najd and Hasa is halophytic in type and dominance of species tends to vary with locality.

(c) Coastal vegetation
Coastal vegetation develops in numerous areas in close proximity to the Arabian Gulf and the Red Sea where atmospheric humidity is relatively high or where precipitation exceeds 100 m

and there is some run-off. Annuals appear in great profusion after rain and the predominance of certain species varies considerably depending on whether the East or West coast is under consideration.

(d) Foothills

In some areas in the foothills where light precipitation occurs and there is some run-off, the vegetative cover in the wadis usually consists of a sparse tree mosaic in conjunction with tussock-type grasses and herbs.

(e) Montane woodland

Except for small areas at high elevations in the Asir mountains, it would appear that montane woodlands are conspicuous by their absence.

Although a relatively wide range of vegetation will thrive under desert conditions, a knowledge of the requirements of plants and the correct placement and maintenance of planting material is of paramount importance in planning landscaping projects in the Middle East.

As this short planting guide attempts to cover a wide range of planting material, it has not been possible to provide detailed descriptions of plants and it is hoped that the brief outlines together with the line drawings will enable readers to recognise plants and their potential in connection with desert landscaping operations.

Like other desert regions, the quality and availability of water and the difficulties associated with aquifer recharge in low rainfall areas will play an important part in future development as reduced availability, a falling watertable and increased salination could restrict development appreciably.

Climate

Generally speaking, the region falls into the zone where hot summers and relatively warm to cool winters are experienced.

(a) Temperature
Normally, at low elevations, winter temperatures, (November-March/April) very occasionally rise above 36° C and minimum temperatures can fall as low as 4° C in some exposed areas and can remain in the vicinity of 11-15° C throughout the cold season.

Prolonged high temperatures are experienced throughout the summer (April/May-October) and maximum temperatures can rise to 45-47° C for short periods and, as haze tends to be minimal, solar radiation is intense.

(b) Relative Humidity
Although it is difficult to generalise, relative humidity is usually in the medium/high range during the winter months, with a slight rise during the months November-January. During the summer months, (April-September/October), relative humidity tends to drop to the low or medium/low range.

(c) Rainfall
The whole region falls into a low rainfall zone, and precipitation is meagre, variable and extremely erratic.

A large proportion of the region has an annual average of less than 30-40 mm and rainfall in the Rub al-Khali is practically non-existant and long periods may elapse between showers.

In the Northwest, the Hejaz, the Western highlands, the Tihama and in the vicinity of the Arabian Gulf, the average rainfall may vary between 70 and 150 mm, and in the Yemeni highlands, the Quarra and other mountainous areas in the South of the region that are influenced by the monsoon, the annual average can rise to over 300 mm.

(d) Wind
Under desert conditions, hot air rises rapidly, and the movement of air currents tend to produce windy conditions and almost continuous high winds can be experienced throughout the year.

In the vicinity of the Arabian Gulf, the Eastern Province and the Najd, the prevailing winds are mainly from the North or Northwest, and to a lesser extent from the Northeast. The mountainous areas in the south of the region are affected to a certain extent by monsoon-type winds, and Westerlies produce some precipitation in the Tihama and the Western coastal plain.

In areas such as Jeddah, wind direction usually varies between North, N.N. West and West, and high winds are not uncommon during the months April/May and December/January. The high, relatively cold, "Shimal" winds that are experienced during the winter months are often accompanied by rain and tend to defoliate crops and deposit saline sand particles on foliage causing "scorch" and "saline burn". Shimals can build up in intensity causing severe, short-term, dust-storms between the months of January/May and, in the Bahrain area, Shimal conditions can be experienced for approximately twenty-three days in June.

Occasionally, hot winds during the summer months tend to "sand-blast" plants, and saline-laden winds in coastal areas can have a deleterious effect on crops and ornamentals, and increase the incidence of chlorosis.

Generally speaking, winds throughout the region tend to produce high evaporation and are, in all probability, one of the greatest limiting factors in the Middle East. The establishment of windbreaks, suitable ground covers and other conservation measures should receive careful consideration in the preliminary stage of any development scheme in the region.

(e) Altitude

Altitude has a significant effect on environmental conditions in desert locations, and even a relatively low altitude of approximately 600 m can prove to be a limiting factor as far as many tropical plants are concerned.

Soils

Soils throughout the region vary considerably, are relatively shallow, and high winds – especially in sandy areas – can cause soil movement on a vast scale.

The spectrum of soil types is extensive and covers coarse, medium and fine sands, gravel formations, sandy/clay loams and sandy loams.

Desert soils are usually alkaline in character, and weathering is characterised by the formation of impermeable or cemented hardpans which are occasionally exposed by wind movement. In areas where sand laminae covers a hardpan, crusting on the surface and accelerated runoff is not uncommon.

Heavy textured silty/clay soils with a platy structure can pose many problems, and although they can retain moisture to a certain extent, permeability is usually reduced appreciably.

Particle size is often exceptionally fine, and in a dry state the surface of the soil tends to break down into a fine dust with traffic; and under irrigation, compaction, panning, surface-crusting, loss of structure, lack of porosity and the retention of salts is not unusual.

Crusting of the surface and panning is also prevalent in areas where clay subsoils restrict drainage in the lower horizon, and fine layers of clay within 30-45 cms of the surface tend to inhibit plant growth and reduce crop yields.

The accumulation of salts in desert soils can pose many problems and may, under certain circumstances, necessitate the use of amendments and specialised techniques. Although sodium concentrations can be relatively high, plant growth may not be inhibited as sodium is often offset to a certain extent by a high calcium value.

The presence of boron can limit the scope and range of crops and ornamentals, and the removal of boron by leaching is usually a long drawnout and expensive proposition.

With carbonates, plant tolerance varies, and high values may enhance "sodium effect". Chloritic symptoms can occur through harmful concentrations of manganese salts, by the unavailability of iron, or by the inability of plants to take up iron under saline conditions.

In view of the difficulties associated with soils throughout the region, the importance of an efficient soil analysis should not be overlooked.

Water potential

The hydrological or natural sources of water throughout the region are rainfall, groundwater and flood flows.

With the exception of a few highland areas, rainfall for direct agricultural use or area recharge is relatively insignificant, and except for brief periods of intense precipitation, there is always a rainfall deficit and evapotranspiration exceeds rainfall appreciably.

Rainfall normally recharges groundwater reservoirs by indirect recharge through runoff in wadis, and although some acquifers have been discovered in the course of oil exploration, the presence of groundwater has in the past been limited to various oases.

Data on the water balance of groundwater bodies, overdrawing and the feasibility of artificial recharge from flood or recycled water is negligible, and it is estimated that an appreciable quantity of flood water is wasted.

Artificial sources of water centre round desalinization and the recycling of treated sewage effluent, and increased desalinization projects may provide sufficient water for future settlements in coastal areas. In the interior, groundwater may be available in some areas, though as most of it is poor quality, it may have to be treated before use, and skilled management will be required if it is used for agricultural purposes over a period of time.

If the maximum use is to be made of recycled effluent, storage will be required to optimise water distribution and meet the variable seasonal irrigation demand, and irrigation systems will have to be designed to minimise losses from bare soil.

Trees

ACACIA
arabica (a)

ALBIZZIA
lebbec (b)

PLATE 1

Latin name	Description
ACACIA *arabica*-	*Plate 1 (a)* Medium sized tree, 7-10 m in height, rounded crown, sharp spines and fragrant yellow flowers; hardy, does well under desert conditions. Suitable as a specimen tree, for background planting and screening. Prefers a dry, well-drained situation. Susceptible under 25°-32° F Propagation: from seed Saline tolerance: excellent Water requirements: regular watering during establishment, light watering during first summer after establishment Wind resistance: excellent
ACACIA *cyanophylla*-	Small tree, 3-4.5 m in height, spreading crown with feathery blue/grey leaves and inconspicuous flowers. Relatively delicate under local conditions, requires a reasonably dry and well-drained situation and is suitable for background planting. Susceptible under 25° F Propagation: from seed or cuttings Saline tolerance: fair Water requirements: regular watering during establishment, light watering after establishment Wind resistance: protection required in early stages
ACACIA *decurrens dealbata*-	Small to medium sized evergreen tree, 6-9 m in height, with greyish green leaves and yellow flowers. Comparatively delicate under local conditions. Suitable for grouping or background planting. Susceptible under 25°-32° F Propagation: from seed or cuttings Saline tolerance: fair Water requirements: regular watering during establishment, light watering after establishment Wind resistance: fair to good
ACACIA *decurrens mollis*-	Small, slender tree, 4-6 m in height with yellowish/green leaves and yellow flowers. Relatively hardy under local conditions. Suitable for specimen planting, background and surround planting. Susceptible under 22° F Propagation: from seed or cuttings Saline tolerance: excellent

AZADIRACHTA
indica (a)

BAUHINIA
alba var candida (b)

BAUHINIA
purpurea (c)

PLATE 2

Water requirements: regular watering during establishment, light watering after establishment
Wind resistance: good

ALBIZZIA *lebbec-*

Plate 1 (b)
Large, deciduous tree, 18-24 m in height with spreading crown; flowers greenish white, fragrant; foliage light. Suitable for large gardens, parks and open spaces. Should not be planted near houses or on avenues as branches are brittle.
Susceptible below 20° F
Propagation: from seed or cuttings
Saline tolerance: reasonably good
Water requirements: regular watering during establishment, light watering at fairly regular intervals after establishment.
Wind resistance: fair, requires protection in early stages

AZADIRACHTA *indica-*

Plate 2 (a)
A small, medium sized tree, 6-8 m in height, evergreen, with a straight stem and small pinnate leaves, bears white, scented flowers, is often used as an avenue tree or for background planting. Is drought resistant and prefers a well-drained situation.
Susceptible below 25° F
Propagation: from seed
Saline tolerance: reasonably good
Water requirements: regular watering during year of establishment, light watering after establishment especially during the summer months
Wind resistance: good, requires protection in early stages

BAUHINIA *alba var candida-*

Plate 2 (b)
A small tree or large bush, 3-4 m in height; bears attractive white flowers, usually blooms while the leaves are dropping or in the event of a cold spell after the leaves have dropped. Suitable for massing or background planting, prefers a well-drained situation.
Susceptible below 25°-32° F
Propagation: by seed
Saline tolerance: fair
Water requirements: regular watering
Wind resistance: fair, requires some protection especially in early stages

BAUHINIA *purpurea-*

Plate 2 (c)
A small tree or large bush, 4-6 m in height, very bushy in growth, bears bunches of mauve/pink orchid-like flowers that appear while the leaves are still on the tree. Suitable for massing or background planting, prefers a well-drained situation.

CALLISTEMON
lanceolatus (a)

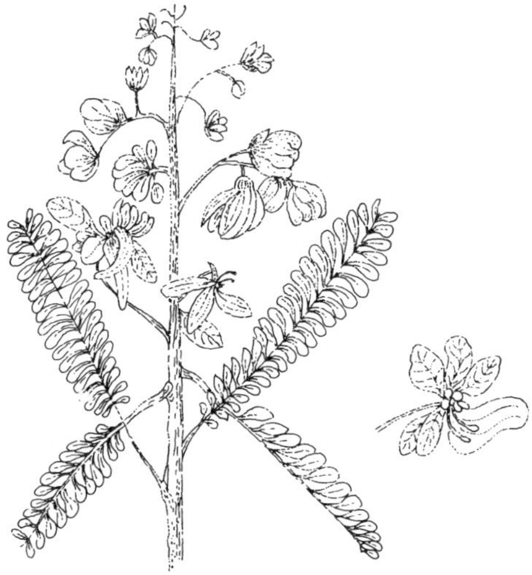

CASSIA
multijuga (c)

CASSIA
fistula (b)

PLATE 3

Susceptible below 25°-32° F
Propagation: by seed
Saline tolerance: fair
Water requirements: regular watering
Wind resistance: fair, requires some protection especially in early stages

BAUHINIA
punctata-

Somewhat similar to *B. purpurea*, bears magenta/white flowers, tends to be a little delicate under M. East conditions especially in the seedling stage.
Susceptible below 25°-32° F
Propagation: by seed
Saline tolerance: poor
Water requirements: regular watering
Wind resistance: poor

BAUINIA
tomentosa-

A small shrub or tree, 2-3 m in height, bearing bright yellow flowers, tends to be a little delicate in seedling stage.
Susceptible below 32° F
Propagation: by seed
Saline tolerance: poor
Water requirements: regular watering
Wind resistance: poor under Middle East conditions

BUTEA
frondosa-

An erect tree with broad, leathery, trifoliate leaves, 6-9 m in height; drought resistant, bears a profusion of crimson or orange/scarlet flowers and is remarkable for its attractive terminal inflorescence. Suitable for background planting and prefers a well-drained situation.
Susceptible below 20° F
Propagation: by seed
Saline tolerance: reasonably good
Water requirements: regular watering during first year of establishment, light watering if and when required after establishment
Wind resistance: excellent, requires some protection in early stages

CALLISTEMON
lanceolatus-

Plate 3 (a)
Large shrub or small tree 2-6 m in height, with oblong, pointed leaves with pink tinge when young. Flowers brick red with long projecting bristles. Hardy, slow growing, prefers dry, well-drained sandy soil and open situation. Suitable for specimen planting, inner-line roadside planting or grouping.

CASUARINA
cunninghamiana (a)

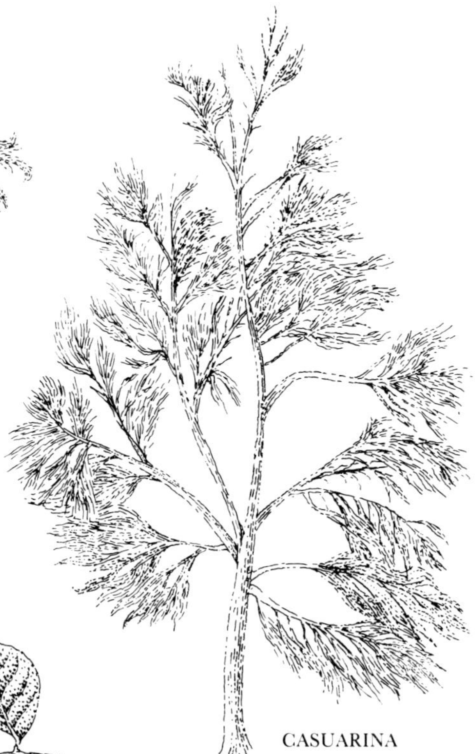

CASUARINA
equisetifolia (c)

CERATONIA
siliqua (b)

PLATE 4

Susceptible below 20°-25° F
Propagation: by seed
Saline tolerance: fair
Water requirements: regular watering
Wind resistance: reasonably good

CASSIA *fistula*-

Plate 3 (b)

A small, attractive flowering tree, 6-9 m in height, with glabrous leaves and large, scented, yellow flowers on drooping racemes. Slow growing, rather delicate under M. East conditions, tends to exhibit chloritic symptoms in early stages, likes full sun and a well-drained soil; suitable for house gardens and parks.
Susceptible below 25°-32° F
Propagation: by seed
Saline tolerance: poor – fair
Water requirements: regular watering
Wind resistance: relatively poor under Middle East conditions

CASSIA *grandis*-

Fairly slow-growing tree, 9-12 m in height; spreading habit, flowers pink fading to coral pink; has large pods and provides good shade. Suitable for large gardens and parks, prefers well-drained situation, slightly delicate under Middle East conditions, tends to exhibit chloritic symptoms in early stages.
Susceptible below 25°-32° F
Propagation: by seed
Saline tolerance: poor – fair
Water requirements: regular watering
Wind resistance: poor, requires protection from wind

CASSIA *multijuga*-

Plate 3 (c)

A small, slender, rather brittle tree 6-7 m in height, very ornamental when in full bloom, bears panicles of bright yellow flowers. Fairly drought tolerant, delicate and difficult in seedling stage. Suitable for small gardens.
Susceptible below 32° F
Propagation: by seed or cuttings
Saline tolerance: fair – poor
Water requirements: regular watering
Wind resistance: poor – fair

CASUARINA *cunninghamiana*-

Plate 4 (a)

A tall, graceful, attractive, quick growing tree 15-18 m in height with fine filliform branchlets instead of leaves; thrives in dry regions; eminently suitable for coastal development and for

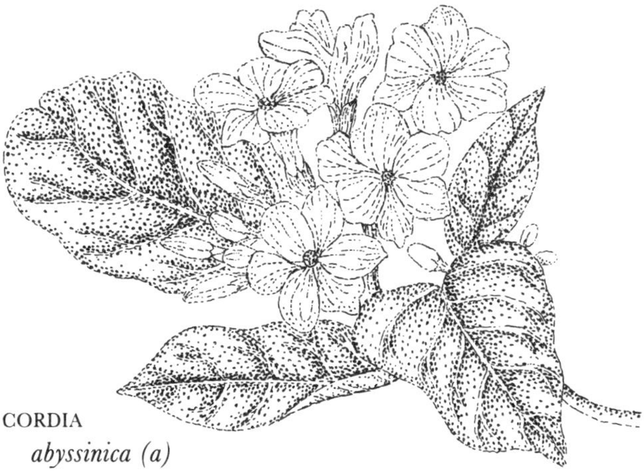

CORDIA
abyssinica (a)

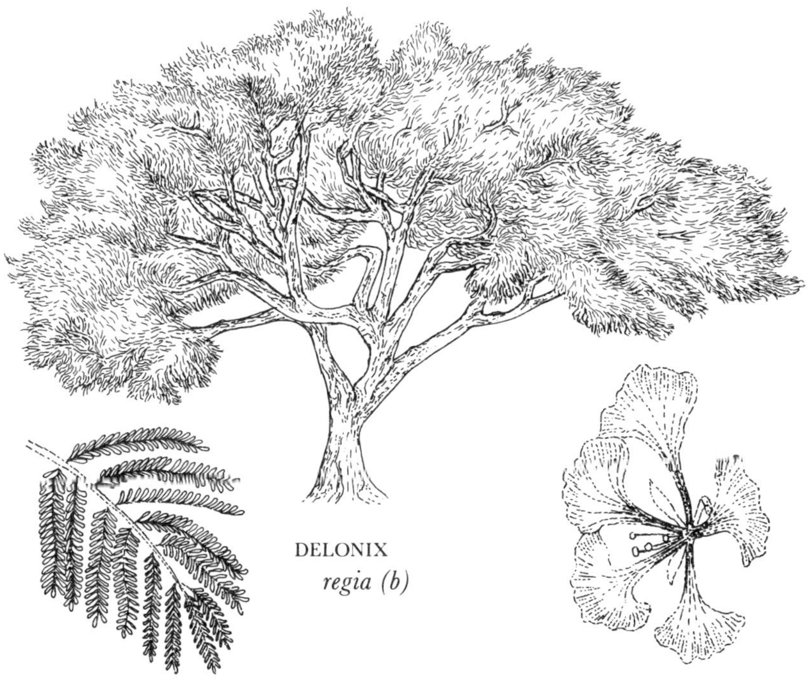

DELONIX
regia (b)

PLATE 5

avenue, background planting and screening, or specimen planting; suitable in large gardens.
Susceptible below 25°-32° F
Propagation: by seed
Saline tolerance: excellent
Water requirements: watering during establishment, light watering after establishment if and when required
Wind resistance: excellent

CASUARINA
equisetifolia-

Plate 4 (c)
A lofty, quick growing tree, 15-21 m in height, somewhat similar to *C. cunninghamiana*, thrives in dry situations, eminently suitable for coastal planting or reclamation, tends to become rather untidy with age. Tolerates cutting back and can form a reasonably good hedge or windbreak. Suitable for large gardens, avenue planting, parks and open spaces.
Susceptible below 25°-32° F
Propagation: by seed
Saline tolerance: excellent
Water requirements: watering during establishment, occasional watering during hot summer weather if required
Wind resistance: excellent

CERATONIA
siliqua-

Plate 4 (b)
A small, spreading tree, 7-9 m in height. Slow growing, of great longevity. Prefers open, well-drained situation.
Susceptible below 20° F
Propagation: by seed
Saline tolerance: excellent
Water requirements: regular watering during establishment, light watering after establishment until trees are mature
Wind resistance: excellent

CORDIA
abyssinica-

Plate 5 (a)
Small, evergreen ornamental tree 5-6 m in height. Produces masses of white flowers. Suitable for background massing or selective planting.
Susceptible below 15° F
Propagation: by seed
Saline tolerance: good to excellent
Water requirements: regular watering during establishment, light watering after establishment
Wind resistance: reasonably good

ERYTHRINA
indica (a)

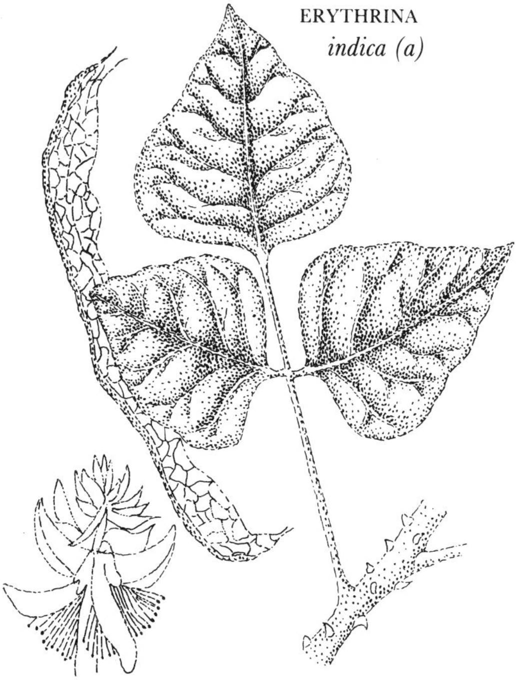

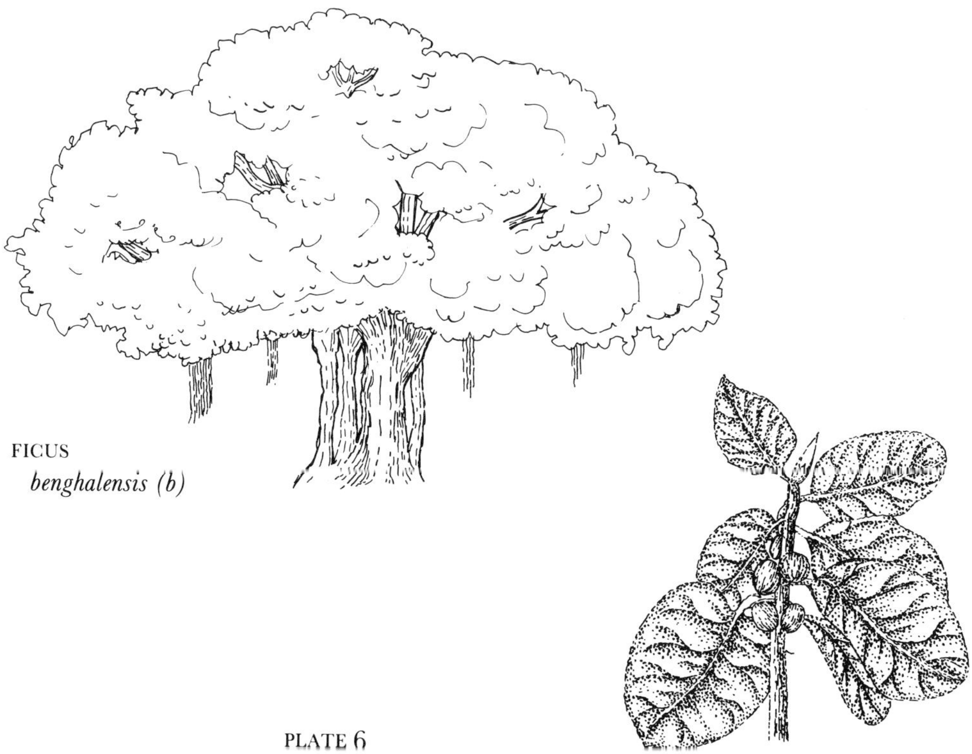

FICUS
benghalensis (b)

PLATE 6

DALBERGIA
sisoo-

Large, deciduous tree, 10-15 m in height, upright in habit with long pinnate, glabrous leaves, suitable for large gardens, avenues and parks.
Susceptible below 20° F
Propagation: by seed
Saline tolerance: reasonably good
Water requirements: watering during establishment, light watering if and when required after establishment
Wind resistance: good

DELONIX
regia-

Plate 5 (b)
A fairly large tree, 12-15 m in height with spreading umbrella-shaped crown, quick growing, flowers red or scarlet; very showy in flower. Suitable for large house gardens, parks, avenues or background planting.
Susceptible below 25°-32° F
Propagation: by seed, cuttings or large branches
Saline tolerance: reasonably good
Water requirements: regular watering during first two years of establishment, light watering at regular intervals after two year period
Wind resistance: fair, requires a well-protected situation

ERYTHRINA
indica-

Plate 6 (a)
Very fast growing, small to medium sized tree, 9-10 m in height, branches and stem armed with thorns which tend to fall off after 2-3 years. Leaves trifoliate, flowers bright red, firm and scentless, are borne in tufts at the tips of branches when they are bare of leaves. Suitable for background planting.
Susceptible below 25°-32° F
Propagation: by seed, cuttings or large branches
Saline tolerance: tolerant to a certain extent
Water requirements: regular watering to attain rapid development
Wind resistance: fair, requires some protection especially in early stages

ERYTHRINA
var indica-

Erythrina lithosperma, E. suberosa, E. variegata etc., all somewhat alike, attractive. Suitable for background planting.
Susceptible below 25°-32° F
E. abyssinaca susceptible below 20° F
Propagation: by seed, cuttings or large branches
Saline tolerance: tolerant to a certain extent
Water requirements: regular watering to attain rapid development

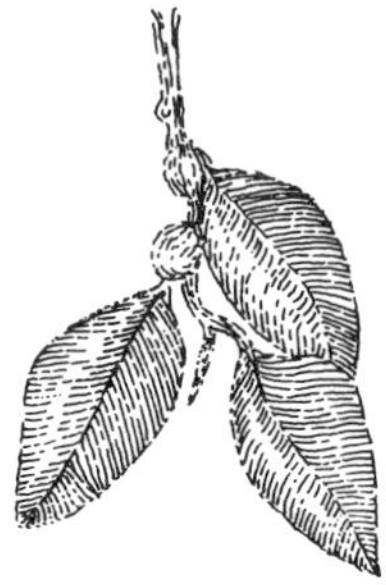

FICUS
benjamina (a)

FICUS
religiosa (b)

PLATE 7

Wind resistance: fair, requires some protection, especially in early stages

EUCALYPTUS *citriodora*-

Handsome, slender tree, 15-21 m in height; prefers fairly dry, well-drained soil. Suitable for avenues, roadside planting, grouping, parks, etc.
Susceptible below 20° F
Propagation: by seed
Saline tolerance: fairly good
Water requirements: regular watering during the first three years after establishment, light watering after three-year period if and when necessary
Wind resistance: protection from wind in early stages improves growth rate

EUCALYPTUS *var*-

E. globulus, *E. ficifolia*, *E. longifolia* etc., varying in height and colour of flowers.
Susceptible below 20° F
Propagation: by seed
Saline tolerance: fairly good
Water requirements: regular watering during the first three years after establishment, light watering after three-year period if and when necessary
Wind resistance: protection from wind in early stages improves growth rate

FICUS *benghalensis*-

Plate 6 (b)
Large, evergreen, slow-growing tree, 12-15 m in height, smooth barked with leathery shiny leaves and bright crimson berries, remarkable for its aerial roots which take root and develop into one immense crown, provides excellent shade, only suitable for large gardens, parks and roadside planting.
Susceptible below 25° F
Propagation: by cuttings
Saline tolerance: reasonably good
Water requirements: regular watering during 2-3 years after establishment, light watering if and when necessary after this period
Wind resistance: protection in early stages improves growth rate

FICUS *benjamina*-

Plate 7 (a)
A large, handsome, spreading tree, 12-15 m in height, small ovate leaves and long, graceful, slender branches; suitable as a shade or avenue tree or specimen planting in large gardens.

GLIRICIDIA
maculata (a)

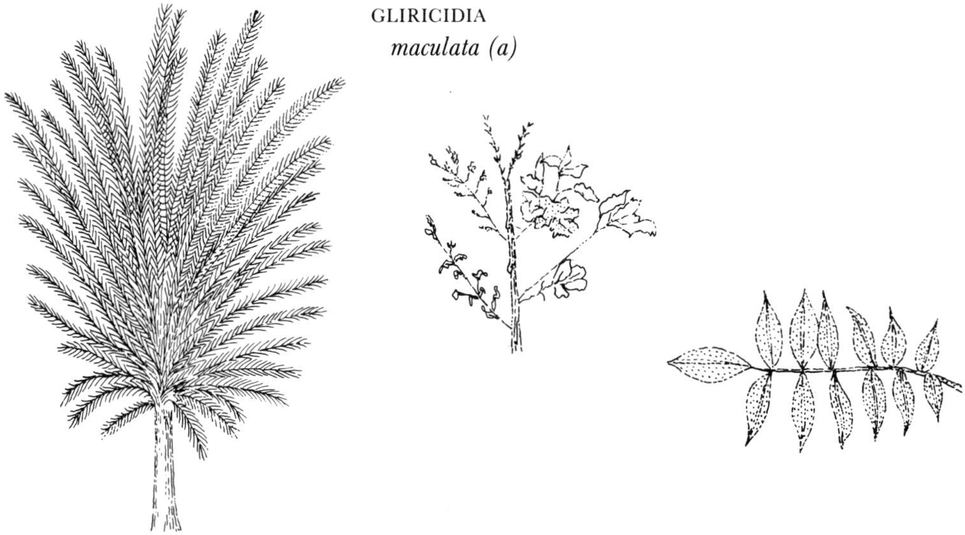

JACARANDA
mimosaefolia (b)

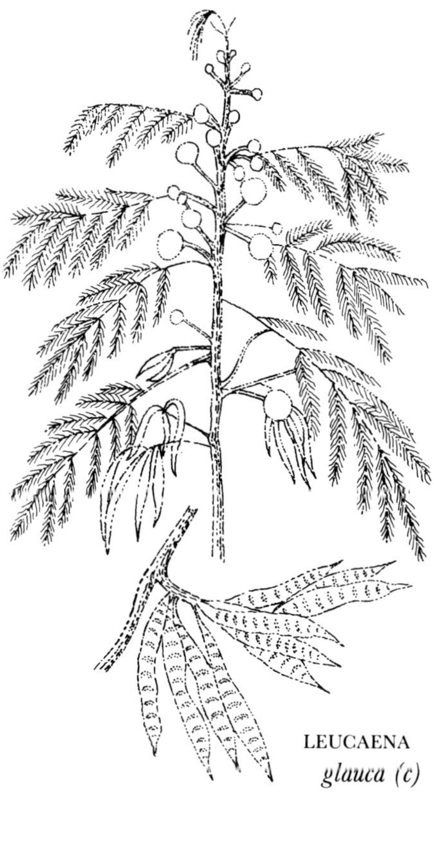

LEUCAENA
glauca (c)

PLATE 8

Susceptible below 25° F
Propagation: by cuttings
Saline tolerance: reasonably good
Water requirements: regular watering during 2-3 years after establishment, light watering if and when necessary after this period
Wind resistance: protection in early stages improves growth rate

FICUS *comosa*-

Somewhat similar to *F. benjamina* in habit of growth, has larger leaves and bright yellow fruit.
Susceptible below 25° F
Propagation: by cuttings
Saline tolerance: reasonably good
Water requirements: regular watering during 2-3 years after establishment, light watering if and when necessary after this period
Wind resistance: protection in early stages improves growth rate

FICUS *religiosa*-

Plate 7 (b)
A large deciduous tree, 18-21 m in height, with an open crown and shiny, drooping leaves, provides light shade, suitable for specimen planting or background planting in large gardens or parks.
Susceptible below 25° F
Propagation: by cuttings
Saline tolerance: reasonably good
Water requirements: regular watering during 2-3 years after establishment, light watering if and when necessary after this period
Wind resistance: protection in early stages improves growth rate

GLIRICIDIA *maculata*-

Plate 8 (a)
A small, quick growing tree, 3-4 m in height, produces long, leafy branches, deciduous, with dense masses of purplish/pink blossoms along the length of the stems, often used as a shade tree for crops; withstands topping at 1½-2½ m. Loppings form an excellent mulch or green manure. Suitable for background planting or screening.
Susceptible below 25°-32° F
Propagation: by seed or cuttings
Saline tolerance: fair

MELIA
azedrach (a)

MILLINGTONIA
hortensis (b)

MORINGA
pterygosperma (c)

PLATE 9

Water requirements: regular watering during establishment, light watering after establishment. Regular watering if toppings are required for mulching
Wind resistance: good. Growth rate is improved if protection is provided in early stage

JACARANDA *mimosaefolia-*

Plate 8 (b)
Small, elegant tree, 7-10 m in height, with attractive feathery leaves and mauve or bluish/purple flowers. Prefers fairly dry situation, well drained soil. Suitable for grouping, avenue or roadside planting on double-lined avenue.
Susceptible below 20° F
Propagation: by seed
Saline tolerance: relatively poor – tends to die back under highly saline conditions
Water requirements: regular watering
Wind resistance: poor; requires protection from wind

LEUCAENA *glauca-*

Plate 8 (c)
Small, quick growing tree, 4-5 m in height, with small, bipinnate leaves which does well in dry districts. Sometimes used as a shade crop; foliage relished by cattle, loppings form an excellent green manure. Suitable for windbreaks, screening and background planting.
Susceptible below 25° F
Propagation: by seed
Saline tolerance: reasonably good
Water requirements: regular watering during establishment, light watering after establishment. Regular watering if loppings are being used for mulching
Wind resistance: reasonably good; protection in early stages increases growth rate

MELIA *azedarach-*

Plate 9 (a)
A small, ornamental tree with bipinnate or tripinnate leaves and small ovoid fruits, native of Iran, 9-12 m in height. Flowers mauve/purple, scented. Suitable for avenue planting (inner line), background planting and for screening. Does exceptionally well under Middle East conditions, fast grower.
Susceptible below 20° F
Propagation: by seed, cuttings or small branches
Saline tolerance: excellent
Water requirements: regular watering during establishment, light watering after establishment
Wind resistance: excellent

PARKINSONIA
aculeata (a)

PISTACHIA
atlantica (b)

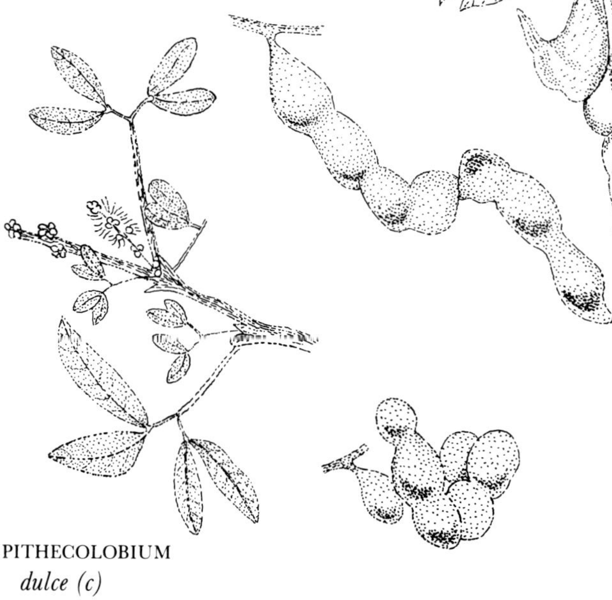

PITHECOLOBIUM
dulce (c)

PLATE 10

MILLINGTONIA *hortensis*-

Plate 9 (b)

A tall, erect tree, 15-18 m in height, with light feathery foliage which bears long, white, tubular, highly scented flowers. Flowers usually appear twice a year in its native habitat; in the Middle East usually flowers only once a year. The wood is rather brittle and the tree tends to break up in high winds. Suitable for avenues or specimen planting.

Susceptible below 20° F

Propagation: by cuttings

Saline tolerance: fair, requires protection from wind

Water requirements: regular watering

Wind resistance: poor

MORINGA *pterygosperma*-

Plate 9 (c)

A short, slender tree, 7-9 m in height with tripinnate leaves and white flowers, ornamental and attractive; prefers a dry, well-drained location. Suitable for specimen planting, background planting and screening.

Susceptible below 25° F

Propagation: by seed

Saline tolerance: fair

Water requirements: regular watering

Wind resistance: fair, requires protection from the wind in early stages of growth

PARKINSONIA *aculeata*-

Plate 10 (a)

A small, quick growing, upright tree, 7-10 m in height, with small leaves, prickly branches and an attractive light green stem. Flowers yellowish/green, suitable for screening, avenue planting (inner line), background planting.

Susceptible below 25° F

Propagation: by seed

Saline tolerance: reasonably good

Water requirements: regular watering during establishment, light watering after establishment

Wind resistance: fair, requires some protection from wind in early stages

PHOENIX *dactylifera*-

A sub-tropical dioecious palm of great antiquity, 18-24 m in height, with handsome pinnate leaves extensively cultivated for its fruit – (date palm). Suitable for avenue, background planting or planting in depth. Hardy and easily propagated.

No susceptibility noticed in Middle East

Propagation: by suckers or offshoots (both male and female)

Saline tolerance: excellent

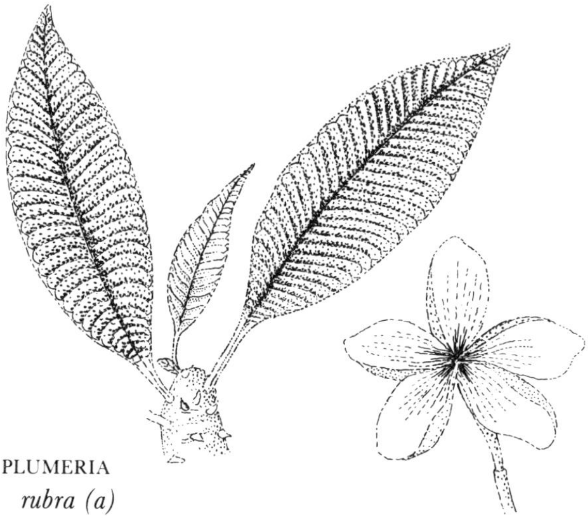

PLUMERIA
rubra (a)

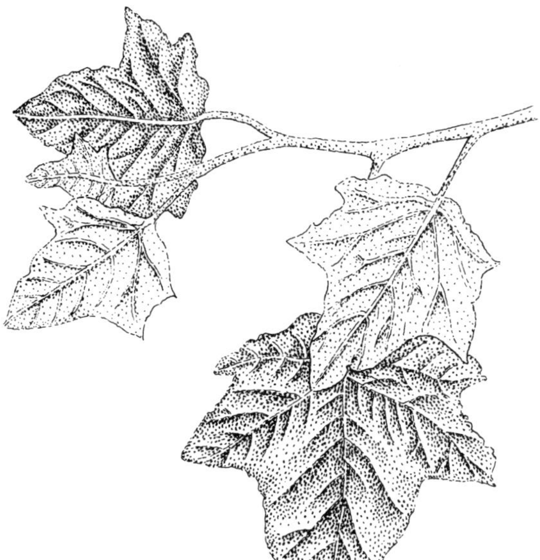

POPULUS
alba (b)

PONGAMIA
glabra (c)

PLATE 11

Water requirements: regular watering if date palms are to be grown for fruit
Wind resistance: excellent

PISTACIA
atlantica-

Plate 10 (b)
Hardy, evergreen tree, 6-10 m in height, upright, attractive foliage. Suitable for specimen and background planting. Grows well in rocky soils in the hills.
Susceptible below 15° F
Propagation: by seed and cuttings
Saline tolerance: good – excellent
Water requirements: regular watering during establishment, light watering – if necessary – after establishment
Wind resistance: excellent

PITHECOLOBIUM
dulce-

Plate 10 (c)
Fairly large, bushy tree, 13-15 m in height, quick growing, provides good shade, eminently suitable for dry, sandy areas. If cut back regularly, forms an impenetrable hedge. In dry areas is suitable for roadside planting, background planting, shade and screening.
Susceptible below 20° F
Propagation: by seed
Saline tolerance: reasonably good
Water requirements: regular watering if rate of growth is to be maintained, or if quick growing hedges are envisaged
Wind resistance: excellent

PLUMERIA
acutifolia-

Small, fleshy, deciduous tree, 4-6 m in height; large, light green leaves, flowers creamy/yellow, waxlike and very fragrant. Suitable for beds, borders, grouping or specimen planting on lawns; does not relish high watertables.
Susceptible below 25° F
Propagation: by cuttings
Saline tolerance: poor
Water requirements: regular watering at frequent intervals
Wind resistance: relatively poor – requires protection from wind at all stages of growth

PLUMERIA
rubra-

Plate 11 (a)
Somewhat similar to *P. acutifolia* with red/crimson flowers. *P. rubra* more robust than *P. acutifolia,* other remarks as for *P. acutifolia.*
Susceptible below 25° F
Propagation: by cuttings
Saline tolerance: relatively poor

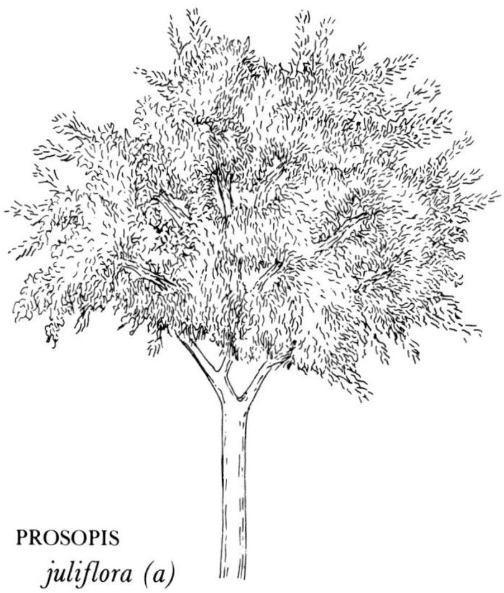

PROSOPIS
juliflora (a)

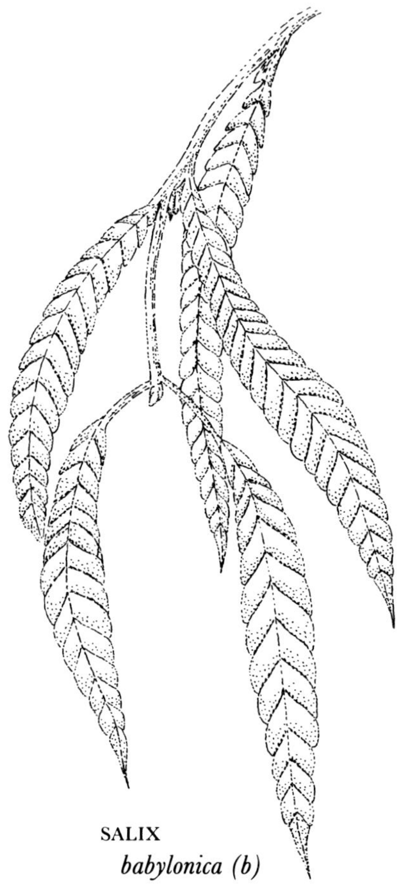

SALIX
babylonica (b)

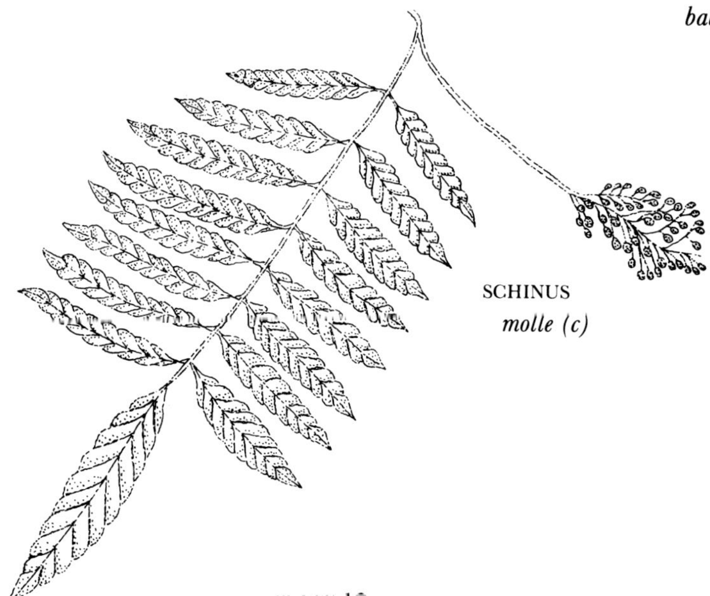

SCHINUS
molle (c)

PLATE 12

Water requirements: regular watering at frequent intervals
Wind resistance: poor – requires protection from wind

PONGAMIA *glabra*-

Plate 11 (c)
A handsome tree, 6-9 m in height with glossy, pinnate leaves bearing racemes of creamy white scented flowers. Prefers a dry, well-drained situation. Suitable for background planting or street planting, or massing in depth.
Susceptible below 20° F
Propagation: by seed
Saline tolerance: fairly poor – especially in early stages of growth
Water requirements: regular watering
Wind resistance: fair

POPULUS *alba*-

Plate 11 (b)
Large, hardy, quick growing deciduous trees, 15-21 m in height. Prefers a fairly moist situation; suitable for avenue planting, windbreaks, screens or specimen planting.
Susceptible below 15° F
Propagation: by seed or cuttings
Saline tolerance: fair
Water requirements: regular watering during establishment, light watering – if necessary – after establishment
Wind resistance: reasonably good

POPULUS *euphratica*-

Large, hardy, quick-growing deciduous trees, approx. 18-24 m in height. Prefers a fairly moist situation; suitable for avenue planting, windbreaks, screens or specimen planting.
Susceptible below 20° F
Propagation: by seed or cuttings
Saline tolerance: good – fair
Water requirements: regular watering during establishment, light watering after establishment
Wind resistance: reasonably good

PROSOPIS *juliflora*-

Plate 12 (a)
A small leguminous tree, 4-6 m in height, leaves, pods and shoots can be used for stock feed. Branches rather brittle, suitable for screening or windbreaks; thrives under dry conditions.
No susceptibility noticed in Middle East
Propagation: by seed
Saline tolerance: excellent
Water requirements: regular watering during establishment, little or no water required after establishment
Wind resistance: excellent

SESBANIA
grandiflora (a)

SPATHODEA
campanulata (b)

PLATE 13

SALIX
babylonica-

Plate 12 (b)
A small to medium sized tree, 6-8 m in height, with handsome drooping foliage and slender branches. Suitable for specimen planting on a lawn.
Susceptible below 20° F
Propagation: by cuttings
Saline tolerance: fair
Water requirements: regular watering at frequent intervals
Wind resistance: fair

SCHINUS
molle-

Plate 12 (c)
A small, spreading tree of ornamental habit, 6-9 m in height, with handsome drooping foliage. Eminently suitable for background planting, screening or specimen planting.
Susceptible below 15°-20° F
Propagation: by seeds or cuttings
Saline tolerance: excellent
Water requirements: regular watering during establishment, light watering after establishment
Wind resistance: good to excellent

SCHINUS
terebinthifolius-

Small, attractive tree, 4-6 m in height, leaves dark green, bears crimson berries in December to January, hardy. Suitable for background planting, screening and planting in depth.
Susceptible below 15°-20° F
Propagation: by seed or cuttings
Saline tolerance: reasonably good
Water requirements: regular watering during establishment, light watering after establishment
Wind resistance: reasonably good

SESBANIA
grandiflora-

Plate 13 (a)
A small, erect tree, quick growing, sparsely branched, soft wooded tree, 4-7 m in height, with an attractive graceful habit. Leaves dark, pinnate, flowers large, pendulous, white or magenta-pink in colour with an unpleasant smell, flowers edible. The tree is short-lived; and does well under Middle East conditions in coastal areas. Suitable for specimen or background planting or screening.
Susceptible below 32° F
Propagation: by seed
Saline tolerance: excellent; also resistant to saline spray
Water requirements: regular watering during establishment
Wind resistance: good to excellent

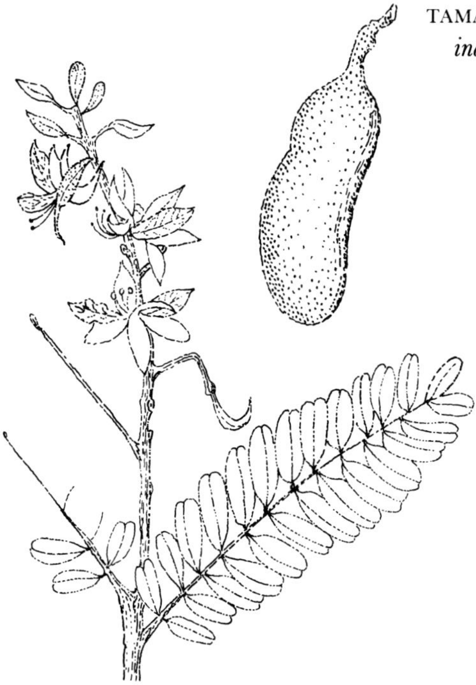

TAMARINDUS
indica (a)

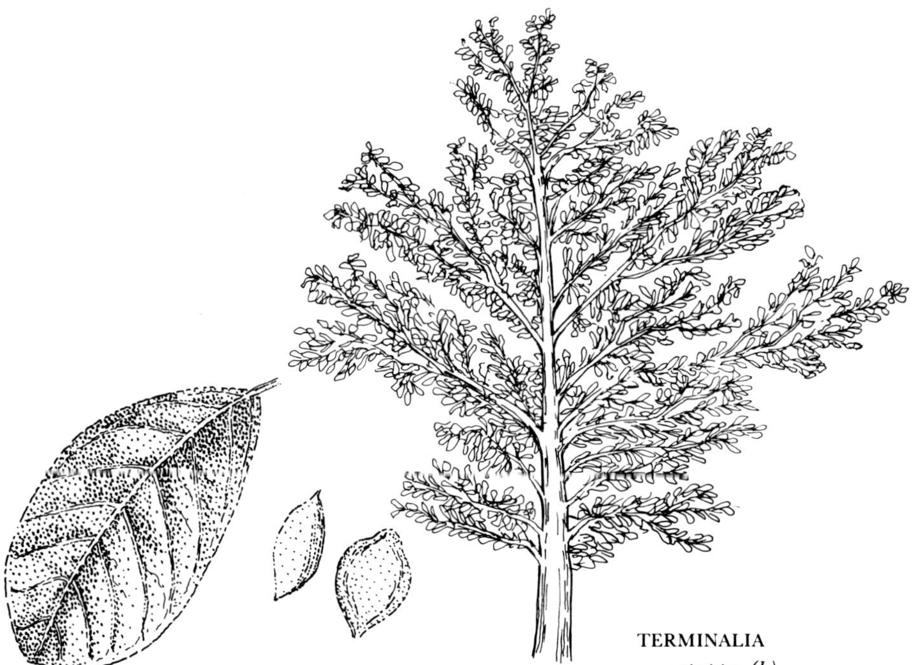

TERMINALIA
catappa (b)

PLATE 11

SPATHODEA
campanulata-

Plate 13 (b)

A medium to tall tree, 12-18 m in height, upright growth, fairly heavy crown, bears very large red, cup-shaped flowers at the tips of the branches, shallow rooted, tends to be blown over by high winds; exhibits chloritic symptoms in early stages. Suitable for large gardens and parks, should not be planted too near houses as old trees tend to become hollow and dangerous.

Susceptible below 25°-32° F

Propagation: from seed, suckers or cuttings

Saline tolerance: fair

Water requirements: regular watering

Wind resistance: fair

SPATHODEA
nilocta-

Somewhat similar to *S. campanulata;* flowers orange/red, brighter than *S. campanulata*. Remarks as for *S. campanulata*.

Susceptible below 25° F

Propagation: from seed, suckers or cuttings

Saline tolerance: fair

Water requirements: regular watering

Wind resistance: fair, requires protection during early stages of growth

TAMARINDUS
indica-

Plate 14 (a)

A large, handsome tree, 15-18 m in height with dense, symmetrical crown and small, pinnate leaves, does well in dry areas. Suitable for roadside planting or where large, attractive trees are required as a background, shade is relatively light, flowers yellow streaked with red; inconspicuous.

Susceptible below 25°-32° F

Propagation: by seed

Saline tolerance: reasonably good

Water requirements: regular watering during establishment, light watering after establishment

Wind resistance: good

TAMARIX
var-

Large bushes or small trees 2-6 m in height, with slender, leafless, twiggy branches, flowers pinkish or cream. Eminently suitable for arid, sandy localities, for beach and desert planting.

Susceptible below 20° F

Propagation: by cuttings

Saline tolerance: excellent

Water requirements: regular watering during establishment, light watering – if required, after establishment

Wind resistance: excellent

THEVETIA
nerefolia (a)

ZIZYPHUS
spina-christi (b)

PLATE 15

TERMINALIA
catappa-

Plate 14 (b)
A fairly tall, deciduous tree, 18-21 m in height; leaves large, branching in whorls, sheds its leaves twice a year. Leaves tend to turn red before winter leaf-drop. Suitable for large house gardens roadside planting, screening and for large open spaces.
Susceptible below 25° F
Propagation: by seed
Saline tolerance: excellent
Water requirements: regular watering during establishment, light watering after establishment
Wind resistance: excellent

THESPESIA
populnea-

A small, hardy tree, 2-3 m in height, the small hibiscus-like flowers open light yellow and turn purple by nightfall. Drought resistant. Suitable for narrow parkway planting, roadside planting and background planting.
Susceptible below 25° F
Propagation: by seed
Saline tolerance: excellent, also resistant to salt spray
Water requirements: regular watering during establishment, light watering – if necessary – after establishment
Wind resistance: excellent

THEVETIA
nerefolia-

Plate 15 (a)
A small, hardy tree, 4-6 m in height, quick grower, flowers profusely. Flowers golden-yellow or orange. Prefers dry, well-drained soil. Suitable for use in groups or singly.
Susceptible below 25°-32° F
Propagation: by seed or cuttings
Saline tolerance: good to excellent
Water requirements: regular watering during establishment, light watering after establishment
Wind resistance: reasonably good

TIPUANA
tipu-

Small/medium evergreen, first growth tree 7-9 m in height with showy bright yellow butterly-shaped flowers. Fast growing tree, suitable for shade and street planting.
Susceptible below 20° F
Propagation: by seed
Saline tolerance: excellent
Water requirements: regular watering during establishment, light watering after establishment
Wind resistance: reasonably good

ZIZYPHUS
spina-christi-

Plate 15 (b)

A small, slow growing deep rooted evergreen, 7-9 m in height. Branches armed with thorns, leaves grey/green. Saline tolerant, hardy and spreading in habit, fruit edible; suitable for screening or wind-breaks.

Susceptible below 20° F

Propagation: by seed or cuttings

Saline tolerance: excellent

Water requirements: regular watering during establishment, light watering after establishment

Wind resistance: excellent

ZIZYPHUS
spina-vulgare-

A small, thorny tree, 4-6 m in height, does well under Middle East conditions. Suitable for windbreaks and screening.

Susceptible below 20° F

Propagation: by seed

Saline tolerance: excellent

Water requirements: regular watering during establishment, light watering after establishment

Wind resistance: excellent

Note: For quick reference see Appendix A
For classification according to colour see Appendix A/1

Flowering and foliage shrubs

ACALYPHA
wilkesiana (a)

ARALIA
balfouri (b)

BOUGAINVILLIA
var (c)

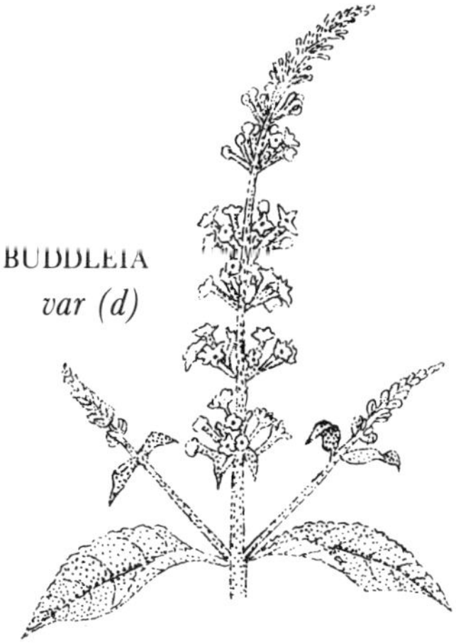

BUDDLEIA
var (d)

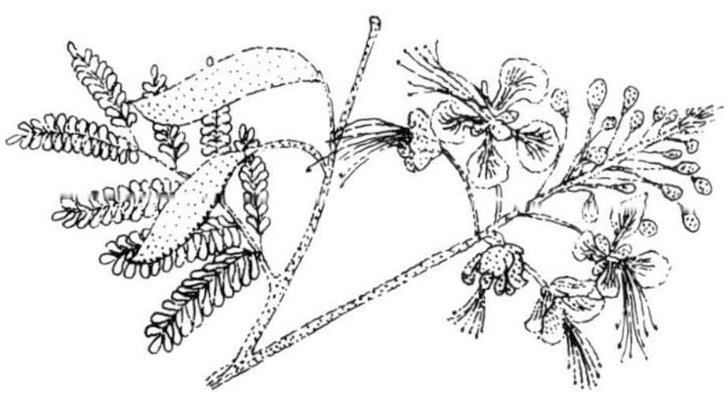

CAESALPINA
pulcherrima (e)

PLATE 16

Latin name	Description
ACACIA *farnesiana-*	A thorny, arborescent shrub, 2½-3 m in height. Flowers yellow, bulb-shaped, crowded on leaf axils. Very sweet smelling. Suitable for screening and background planting or grouping. Susceptible below 22° F Propagation: by seed or cuttings Saline tolerance: reasonably good Water requirements: regular watering during establishment, light watering after establishment if necessary Wind resistance: good to excellent
ACALYPHA *wilkesiana-*	*Plate 16 (a)* Bushy, ornamental shrub, 1-3 m in height with dark green or red stems and variegated leaves varying in colour from grey/green to brownish-red. Hardy, drought resistant, quick grower, suitable for borders, hedges and background planting in beds. Responds well to pruning. Susceptible below 25 ° F Propagation: by cuttings Saline tolerance: reasonably good Water requirements: regular watering Wind resistance: good Other varieties that do well under M. East conditions are *A. illustris, A. marginata, A. triumphans.*
ARALIA *balfouri-*	*Plate 16 (b)* A small, bushy shrub, 2-2½ m in height with pinnate, variegated leaves, leaflets oval or circular, freely blotched with grey or white. Prefers a semi-shaded situation. Suitable for specimen or background planting, can also form an attractive hedge. Susceptible below 25° F Propagation: by cuttings Saline tolerance: fair Water requirements: regular watering Wind resistance: poor, requires protection from winds
ATRIPLEX *halimus-*	Small, perennial, salt-bush 1-1½ m in height. Thrives in arid areas on dry, sandy soils, hardy, drought resistant, leaves silvery grey in colour. Suitable for low hedges, border planting or massing in depth on bare open spaces. Susceptible below 20° F Propagation: by cuttings Saline tolerance: excellent Boron tolerance: excellent

Water requirements: light watering during establishment, very occasional watering after establishment in the event of exceptionally dry conditions
Wind resistance: excellent

BOUGAINVILLIA
formosa-

Plate 16 (c)
A variety of the Glabra group of Bourgainvillias 1½-2 m in height, grown as a shrub on a support or as a standard. Very attractive mauve/purple bracts, does well under Middle East conditions. Tends to exhibit chloritic symptoms in early stages. Prefers well drained soil and fairly sunny situation and some protection from wind.
Susceptible below 25°-32° F
Propagation: by cuttings
Saline tolerance: fair
Water requirements: regular watering
Wind resistance: poor, requires protection from wind

BUDDLEIA
alternifolia-

Large shrub, 3-3½ m in height, attractive lilac flowers borne on racemes on the end of twigs. Exhibits chloritic symptoms in all stages of growth, slightly delicate under Middle East conditions. Suitable for grouping or background planting, requires some protection.
Susceptible below 25°-32° F
Propagation: by seed or cuttings
Saline tolerance: poor
Water requirements: regular watering
Wind resistance: poor, requires protection from wind

BUDDLEIA
asiatica-

Large shrub 1-2 m in height, leaves lanceolate, pointed, serrated, shiny above and covered with white 'felt' underneath. Flowers small, borne in pendant racemes. Requires a dry, well-drained situation. Suitable for grouping or background planting.
Susceptible below 25° F
Propagation: by seed or cuttings
Saline tolerance: fair
Water requirements: regular watering
Wind resistance: poor, requires protection from wind

BUDDLEIA
lindeyana-

Fairly large shrub, 1-2½ m in height. Flowers elliptical, ovate, mauve/purple in colour borne on long racemes. Exhibits chloritic symptoms in all stages of growth. Requires dry, well-drained situation, suitable for grouping or background planting

Susceptible below 25° F
Propagation: by cuttings
Saline tolerance: poor
Water requirements: regular watering
Wind resistance: poor, requires protection from wind

BUDDLEIA *variablis*-

Plate 16 (d)
Large shrub, 1-3 m in height with a spread of approx. 2-3 m. Leaves pointed, serrated, shiny above and silver/grey beneath. Flowers light purple with orange throat, borne on long racemes. Growth under Middle East conditions slow; exhibits chloritic symptoms throughout growth and requires protection. Suitable for grouping or background planting.
Susceptible below 25°-32° F
Propagation: by cuttings
Saline tolerance: poor
Water requirements: regular watering
Wind resistance: poor, requires protection from wind

BARLERIA *prionitis*-

A thorny, many branched shrub, ½-1½ m in height, leaves narrow and long, flowers sessile, yellow in colour. Prefers dry, well-drained situation. Suitable for hedges, grouping or planting in beds.
Susceptible below 25°-32° F
Propagation: by cuttings
Saline tolerance: fair
Water requirements: regular watering during establishment
Wind resistance: fair, requires some protection against wind

CAESALPINA *pulcherrima*-

Plate 16 (e)
An unbranched, arborescent shrub, 2-4½ m in height. Flowers on and off throughout the year. Leaves bipinnate, flowers bright red with yellow margins that later turn dark red. Rapid growth, responds well to pruning or clipping. Hardy under Middle East conditions. Suitable for ornamental hedges, background and specimen planting.
Susceptible below 25°-32° F
Propagation: by seed, cuttings or layers
Saline tolerance: reasonably good
Water requirements: regular watering
Wind resistance: fairly good, requires some protection from wind in early stages

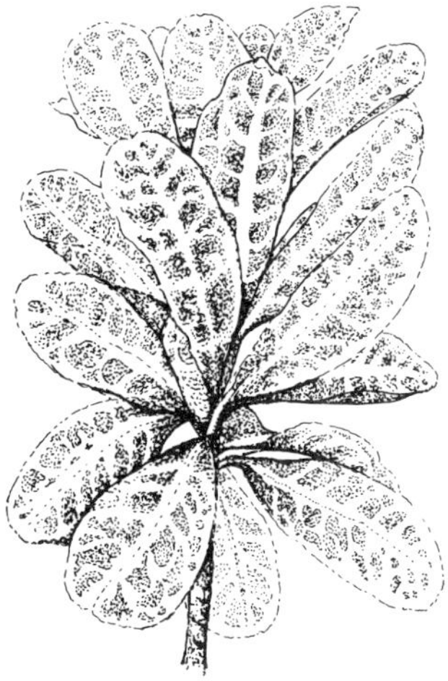

CODIAEUM
var (a)

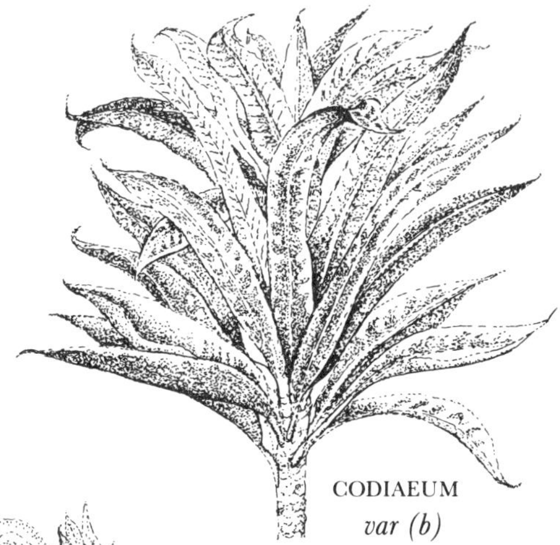

CODIAEUM
var (b)

DATURA
suaveolens (c)

DURANTA
repens (d)

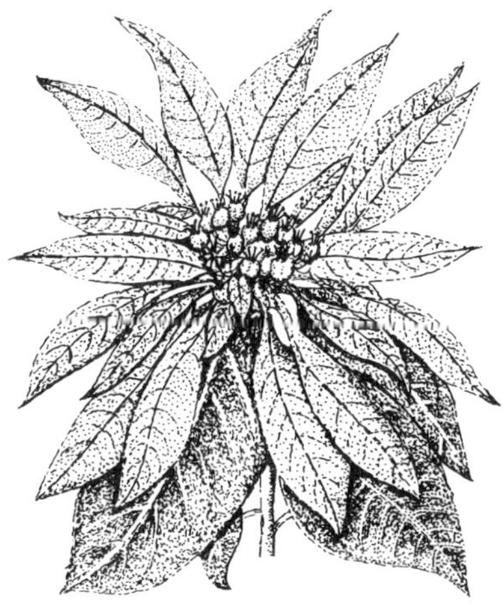

EUPHORBIA
pulcherrima (e)

PLATE 17

CAESALPINA
gillesi-

A large, rather straggling shrub, 2-3 m in height with yellow flowers and bright red stamens. Flowers on and off throughout the year, relatively drought resistant. Suitable for borders, background planting and specimen planting if pruned regularly.
Susceptible below 25°-32° F
Propagation: by seed or cuttings
Saline tolerance: fair to reasonably good
Water requirements: regular watering
Wind resistance: poor, requires protection from wind especially in early stages

CODIAEUM
var-

Plate 17 (a & b)
Large shrubs, 1½-3 m in height, numerous varieties, highly ornamental, leaves in a variety of colours and shapes. Prefers a semi-shaded situation. Suitable for hedges, borders and background planting also for grouping on a lawn.
Susceptible below 25°-32° F
Propagation: by cuttings
Saline tolerance: fair
Water requirements: regular watering
Wind resistance: poor, requires protection from wind

DATURA
fastuosa-

Small, annual shrub, ½-1½ m in height with large scattered, angular leaves and purple or white flowers; flowers isolated, erect. Prefers a dry, well-drained situation in the open. Suitable for specimen planting or grouping. Seeds very poisonous.
Susceptible below 25°-32° F
Propagation: by seed
Saline tolerance: fair
Water requirements: regular watering during establishment, light watering after establishment
Wind resistance: poor, requires protection from wind

DATURA
suaveolens-

Plate 17 (c)
Large shrub, 2-3½ m in height; somewhat similar to *D. fastuosa*, leaves and flowers scattered, white, scented. Requires a dry well-drained, sunny situation. Tends to exhibit chloritic symptoms in early stages. Suitable for specimen planting or grouping. Seeds very poisonous.
Susceptible below 25°-32° F
Propagation: by seed
Saline tolerance: fair
Water requirements: regular watering during establishment, light watering after establishment
Wind resistance: poor, requires protection from wind

DATURA
natalensis-

Fairly large shrub, 1½-2 m in height; somewhat similar to *D. fastuosa*, flowers white or cream, pendulous. Suitable for grouping or background planting. Seeds very poisonous.
Susceptible below 25°-32° F
Propagation: by seed or cuttings
Saline tolerance: fair
Water requirements: regular watering during establishment, light watering after establishment
Wind resistance: poor, requires protection from wind

DODONAEA
viscosa-

Small, bush/tree, 3-3½ m in height, young shoots viscous, flowers greenish/yellow, insignificant. Drought resistant, responds to pruning. Suitable for hedging or screening.
Susceptible below 20°-25° F
Propagation: by seed
Saline tolerance: excellent
Water requirements: regular watering during establishment, light watering if and when required after establishment
Wind resistance: excellent

DURANTA
repens-

Plate 17 (d)
Tall, erect, strongly branched, arborescent shrub, 1-4½ m in height, with square pendant twigs. Leaves small, flowers borne on one side of the stalk, blue/purple in colour; fruits ovate, bulb-shaped, shiny, orange in colour. Plants bloom profusely on and off throughout the year. Prefers a dry, well-drained, open situation; eminently suitable for grouping, specimen planting or for background planting and hedging; responds well to pruning.
Susceptible below 20° F
Propagation: by seed or cuttings
Saline tolerance: reasonably good
Water requirements: regular watering
Wind resistance: reasonably good, requires some protection from wind in early stages

EUPHORBIA
pulcherrima-

Plate 17 (e)
Ornamental shrub, 1-4½ m in height, leaves scattered, ovate-oblong or elliptical. Flowers small, insignificant, bracts vermillion, pink or white. Quick grower, tends to be a little difficult under Middle East conditions and exhibits chloritic symptoms in early stages. Water consumption relatively high during the growing stage. Suitable for background planting or specimen planting or screening.
Susceptible below 25° F

Propagation: by cuttings
Saline tolerance: relatively poor
Water requirements: high, regular watering
Wind resistance: poor, requires protection from wind

HIBISCUS *rosa-sinensis-*

Plate 18 (a)
Very attractive, shapely shrub, 2-3½ m in height, leaves ovate-cordate, coarsely serrated; flowers large and long stalked, yellow, orange, pink, red or purple with dark spot at the base of the lobes. Flowers either single or double, numerous hybrids. Requires a dry, well-drained, sunny situation with some protection from wind. *H. rosa-sinensis* is a "marker plant" and exhibits chloritic symptoms; does not do well in areas with a high water-table. Suitable for beds, borders, as specimen plants, or for ornamental hedges.
Susceptible below 25°-32° F
Propagation: by cuttings, grafting or marcots
Saline tolerance: reasonably good
Water requirements: regular watering
Wind resistance: poor, requires protection from wind

HIBISCUS *schizopetalus-*

Plate 18 (b)
Attractive shrub, 2-3½ m in height, strongly branched with long, slender, reflexed branches. Leaves short stalked, ovate, coarsely serrated, flowers pendant, light red on outside and dark red inside. Prefers a dry, well-drained, open situation with some protection; also does well in partial shade. Like *H. rosa-sinensis*, *H. schizopetalus* is a "marker plant" which can exhibit chloritic symptoms. Eminently suitable in beds, borders, as a specimen plant or for ornamental hedges.
Susceptible below 25°-32° F
Propagation: by cuttings
Saline tolerance: reasonably good
Water requirements: regular watering
Wind resistance: requires protection from wind

IXORA *coccinea-*

Plate 18 (c)
Relatively large shrub, 2-3½ m in height, leaves elliptical, dark green above and light green underneath. Flower trusses small, orange, scarlet or salmon yellow. Can exhibit some chloritic symptoms in early stages, prefers a dry, well-drained soil, requires some protection from wind; suitable for grouping or specimen planting.
Susceptible below 25°-32° F
Propagation: by cuttings
Saline tolerance: poor to fair

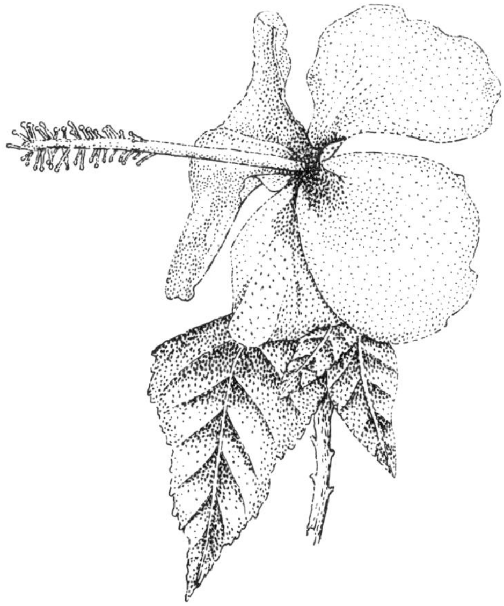

HIBISCUS
rosa-sinensis (a)

HIBISCUS
schizopetalus (b)

IXORA
chinensis (c)

LANTANA
var (d)

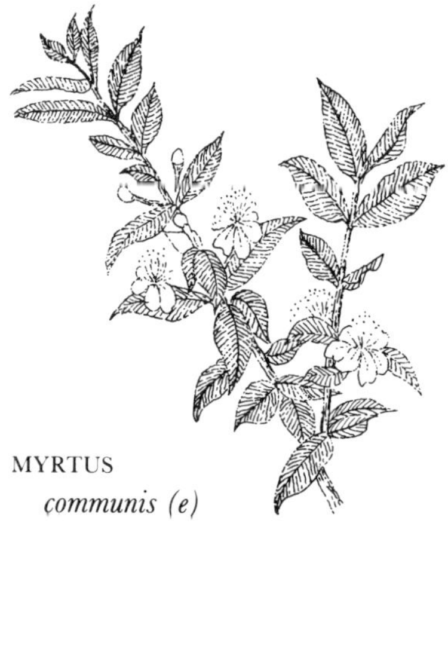

MYRTUS
communis (e)

PLATE 18

Water requirements: regular watering
Wind resistance: poor, requires protection from wind

IXORA
chinensis-

Fairly small, erect shrub, 1-2 m in height, leaves leathery, oblong, flowers scarlet or reddish-orange. Can exhibit chloritic symptoms in early stages of growth. Requires a dry, well-drained situation, can do well in partial shade, slow grower. Suitable for beds, grouping or background planting, or specimen planting.
Susceptible below 25°-32° F
Propagation: by cuttings
Saline tolerance: poor to fair
Water requirements: regular watering
Wind resistance: poor, requires protection from wind

IXORA
javanica-

Large shrub, 2-3½ m in height, with oblong, lanceolate leaves, flowers orange turning to red, borne in terminal trusses. Can exhibit chloritic symptoms and requires a dry, well-drained situation. Suitable for specimen planting or grouping. Slow grower.
Susceptible below 25°-32° F
Propagation: by cuttings
Saline tolerance: poor to fair
Water requirements: regular watering
Wind resistance: poor, requires protection from wind

LANTANA
var-

Plate 18 (d)
Relatively small to very large shrubs, 1 3½ m in height, numerous varieties and hybrids. Only free-flowering varieties are suitable for gardens. Flowers range from white, yellow, pink, orange and red; young flowers are often different in colour from old flowers, giving a two-tone effect. Dwarf mauve and yellow varieties are useful as ground covers or low hedges. Larger varieties are more suitable for hedges, background planting and massing. All types require regular maintenance, as otherwise they tend to become straggly. Lantanas prefer a well-drained, open situation.
Susceptible below 25°-32° F
Propagation: by seed or cuttings
Saline tolerance: reasonably good
Water requirements: regular watering during establishment, light watering after establishment
Wind resistance: reasonably good, requires some protection in early stages

NERIUM
oleander — double (a)

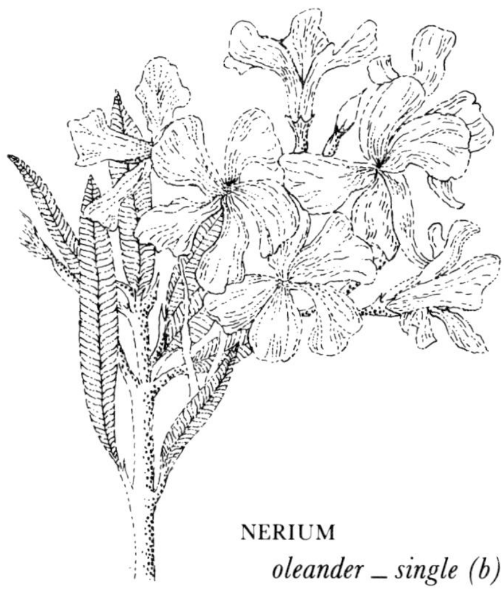

NERIUM
oleander — single (b)

PLUMBAGO
indica (c)

PLATE 19

LAWSONIA
inermis-

A large, close-growing shrub, 2-2½ m in height with slender branches and small, oval, greyish/green leaves. Flowers very highly scented, pink/white in stiff, terminal panicles; fruits globular, green to red in colour. Hardy, quick growing, prefers an open situation. Suitable for hedges and group planting.
Susceptible below 25°-32° F
Propagation: by seed or cuttings
Saline tolerance: excellent
Water requirements: regular watering during establishment, light watering after establishment
Wind resistance: reasonably good, requires some protection from wind in early stages

MYRTUS
communis-

Plate 18 (e)
A large, close-growing shrub, 2-2½ m in height, with small dark green leaves, blue/black berries and small white flowers. The whole plant is fragrant, hardy under Middle East conditions, and prefers a dry, well-drained situation. Eminently suitable for border planting, grouping or hedging.
Susceptible below 25° F
Propagation: by seed or cuttings
Saline tolerance: excellent
Water requirements: regular watering during establishment, light watering after establishment
Wind resistance: reasonably good, requires protection from wind in early stages

NERIUM
oleander-

Plate 19 (a & b)
A large, attractive, free-flowering shrub, 2-3 m in height with narrow, oblong-lanceolate leaves. Prefers a dry, well-drained, open situation. Several single and double varieties, flowers in clusters; colour range covers pink, red, white, yellow and striped. Suitable for back of borders, beds, screening, specimen planting or avenue planting when grown as standards. All varieties respond well to pruning; milky juice and roots poisonous.
Susceptible below 25° F
Propagation: by seed or cuttings
Saline tolerance: good to excellent
Water requirements: high, regular watering
Wind resistance: excellent

OCIMUM
basilicum-

Hardy herb, ½-1 m in height; small leaves and purple flowers. Prefers semi-shaded situation. Suitable in borders or planted along water channels.
Susceptible below 25° F

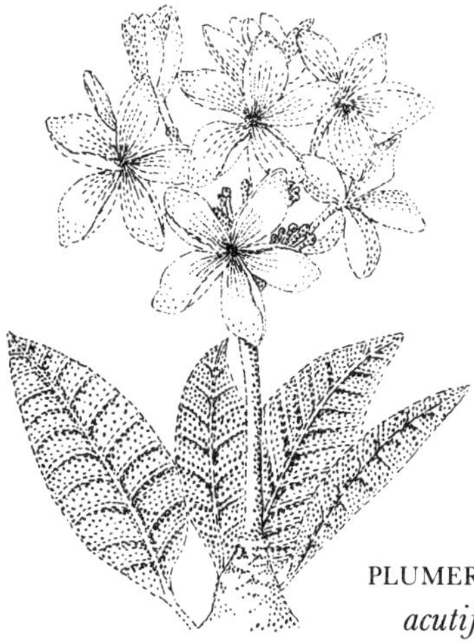

PLUMERIA
acutifolia (a)

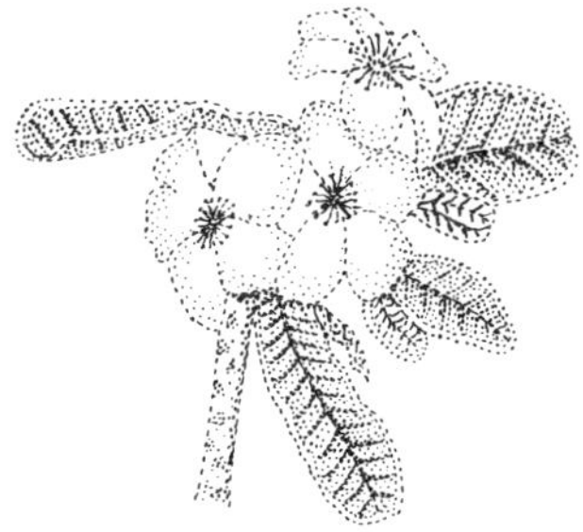

PLUMERIA
obtusa (b)

TECOMA
stans (d)

TECOMARIA
capensis (e)

PLUMBAGO
capensis (c)

PLATE 20

Propagation: by seed
Saline tolerance: reasonably good
Water requirements: regular watering
Wind resistance: poor – fair; requires protection from wind

PLUMBAGO
capensis-

Plate 20 (c)
A slender-stemmed, rambling shrub, ½-1½ m in height with small leaves and trusses of sky-blue, phlox-like flowers. Relatively delicate under Middle East conditions. Tends to exhibit chloritic symptoms. Requires a dry, well-drained, semi-shaded situation with some protection. Suitable for low hedges if supported, or for planting against a wall or bank.
Susceptible below 40° F; tends to drop its leaves below 40°-45° F
Propagation: by seed or cuttings
Saline tolerance: poor
Water requirements: regular watering
Wind resistance: poor, requires protection from wind

PLUMBAGO
indica-

Plate 19 (c)
Similar in some respects to *P. capensis*, ½-1 m in height, flowers pinkish/red to brick-red.
Susceptible below 40° F; tends to drop its leaves below 40°-45° F
Propagation: by seed or cuttings
Saline tolerance: poor
Water requirements: regular watering
Wind resistance: poor, requires protection from wind

PLUMERIA
acutifolia-

Plate 20 (a)
Covered under TREES

PLUMERIA
rubra-

Covered under TREES *(Plate 11 (a))*

PUNICA
granatum-

Plate 21 (a)
Small, ornamental tree/shrub, 2½-3½ m in height, leaves opposite, often clustered on arrested branchlets, lanceolate. Flowers terminal, relatively large, bright red or orange-yellow in colour. Drought resistant, does well under Middle East conditions. Suitable for specimen planting, background planting or screening.
Susceptible below 20° F
Propagation: by seed, layering or grafting
Saline tolerance: good to excellent
Water requirements: regular watering
Wind resistance: excellent

PUNICA
granatum (a)

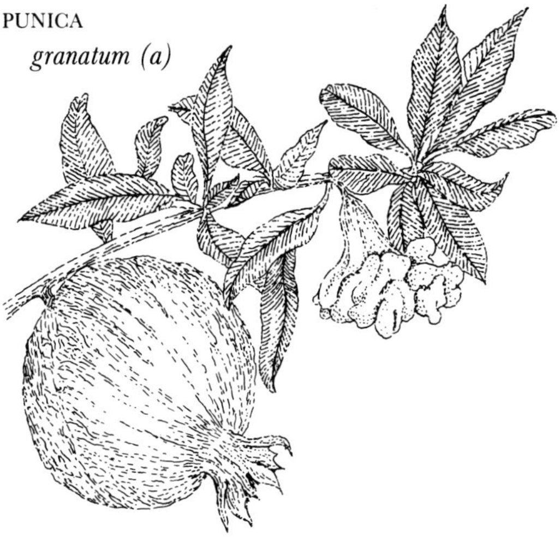

RUSSELIA
juncea (b)

YUCCA
filamentosa (c)

PLATE 21

RUSSELIA
juncea-

Plate 21 (b)
Herbaceous, branched, bare shrub 1-1½ m in height, with rush-like twigs. Leaves oblong, lanceolate, flowers on stalks in loose racemes, scarlet in colour. Flowers profusely on and off throughout the year. Suitable for bedding or grouping.
Susceptible below 25°-32° F
Propagation: by seed and cuttings
Saline tolerance: reasonably good
Water requirements: regular watering
Wind resistance: poor, requires protection from wind

SESBANIA
aegyptiaca-

A tall, quick-growing upright annual/biennial, 2½-4½ m in height with pale, yellow flowers. Suitable for screening or light windbreaks; can be planted for green manure and turned in at ½ m.
Susceptible below 25°-32° F
Propagation: by seed
Saline tolerance: excellent
Water requirements: regular watering if grown for screening or as a green manure crop
Wind resistance: excellent

TECOMA
stans-

Plate 20 (d)
A profusely flowering, branching shrub, 2½-4½ m in height with bunches of yellow, tubular flowers and smooth, pinnate, serrated leaves. The plant flowers on and off throughout the year, does well under Middle East conditions and responds to pruning. Suitable for screening, background planting and grouping. Prefers a dry, well-drained soil and does well in an open situation.
Susceptible below 25°-32° F
Propagation: by seed or cuttings
Saline tolerance: good
Water requirements: regular watering
Wind resistance: reasonably good, requires some protection in early stages

TECOMA
smithii-

Somewhat similar to *T. stans*, 2-2½ m in height, flowers orange tinged with red. Delicate under Middle East conditions, tends to exhibit chloritic symptoms. Suitable for screening, background planting and grouping. Prefers dry, well-drained soil and an open situation.
Susceptible below 32° F
Propagation: by seed or cuttings

Saline tolerance: relatively poor
Water requirements: regular watering
Wind resistance: requires protection from wind

TECOMARIA *capensis*-

Plate 20 (e)
Shrub-climber, 1-3 m in height; leaves ovate, pointed, toothed at the top of the leaflets. Flowers orange/scarlet in terminal racemes with protruding stamens. Flowers well over long periods; exhibits chloritic symptoms. Requires a dry, well-drained, open or semi-shaded situation. Suitable for specimen planting or hedging on supports, or planting along a wall. Delicate under Middle East conditions.
Susceptible below 25°-32° F
Propagation: by seed or cuttings
Saline tolerance: relatively poor
Water requirements: regular watering
Wind resistance: fair, requires some protection

YUCCA *filamentosa*-

Plate 21 (c)
A nearly stemless shrub, 1½-4½ m in height, with stiff, pointed leaves with curly threads on the leaf margins. The flowers are very spectacular, creamy-white in very large panicles. Slow growing, does well under Middle East conditions. Prefers a dry, well-drained, open situation.
Susceptible below 25° F
Propagation: by suckers
Saline tolerance: excellent
Water requirements: regular watering during establishment, light watering after establishment
Wind resistance: excellent

YUCCA *gloriosa*-

Similar in some respects to *Y. filamentosa*, 1½-4½ m in height, leaves edged with brown, flower panicles are yellowish-white. Slow growing, does well under Middle East conditions. Prefers a dry, well-drained, open situation.
Susceptible below 25° F
Propagation: by suckers
Saline tolerance: excellent
Water requirements: regular watering during establishment, light watering after establishment
Wind resistance: excellent

Note: For quick reference see Appendix B
For classification according to colour, see Appendix B/1.

Climbers and creepers

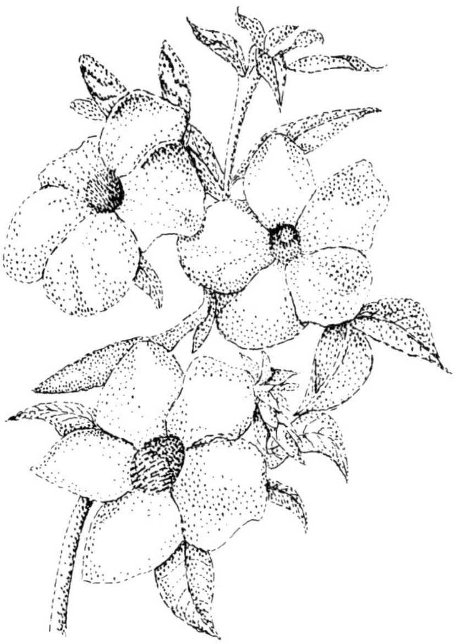

ALLAMANDA
cathartica (a)

ANTIGONON
leptopus (b)

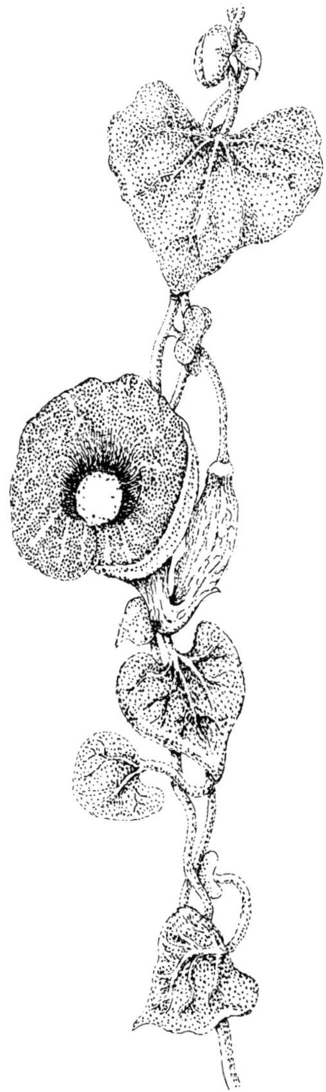

ARISTOLOCHIA
elegans (c)

PLATE 22

Latin name	Description
ALLAMANDA *cathartica-*	*Plate 22 (a)* A handsome climbing shrub with relatively weak and pendant branches, leaves shiny in whorls of four. Flowers large, tubular, golden-yellow in colour, fruit in large spiny pods. Tends to flower on and off throughout the year; exhibits chloritic symptoms, especially in early stages of growth. Prefers a damp, semi-shaded situation and should be 'pinched back' to produce a branching habit. Suitable for training on supports. Susceptible below 32° F Propagation: by cuttings Saline tolerance: poor Water requirements: regular watering Wind resistance: poor, requires protection from wind
ALLAMANDA *hendersonii-*	An allied species to *A. cathartica,* slower growing and more robust. Flowers large, tubular, golden-yellow and fruit in large spiny pods. Also tends to flower on and off throughout the year, exhibiting chloritic symptoms especially in early stages. Prefers damp, semi-shaded site. Suitable for training on supports. Susceptible below 32° F Propagation: by cuttings Saline tolerance: poor Water requirements: regular watering Wind resistance: poor, requires protection
ALLAMANDA *nerifolia-*	Somewhat similar to *A. cathartica,* leaves oblong-ovate in whorls of 2-5, flowers dark yellow with a brown centre. Susceptible below 32° F Propagation: by cuttings and can also be grown from seed Saline tolerance: poor Water requirements: regular watering Wind resistance: poor, requires protection
ANTIGONON *leptopus-*	*Plate 22 (b)* Ornamental climber 9-12 m in length with pointed heart-shaped leaves, flowers white or pinkish-red. Profuse bloomer; prefers a dry, well-drained, fairly open situation and some protection from wind. Suitable for growing on banks, trellises, fences, arbours etc. Susceptible below 25°-30° F Propagation: by seeds, tubers or cuttings Saline tolerance: fair

Water requirements: regular watering
Wind resistance: fair, requires some protection from wind, especially in early stages

ARISTOLOCHIA
elegans-

Plate 22 (c)
Ornamental climber, 4-4½ m in length with large leaves and flowers, brownish/purple with white spots. The flowers tend to give off an unpleasant scent. Tends to exhibit chloritic symptoms in early stages. Prefers a rich, well-drained soil and a fairly sunny protected situation.
Susceptible below 25°-32° F
Propagation: by seed, cuttings or layers
Saline tolerance: poor
Water requirements: regular watering
Wind resistance: poor, requires protection from wind
Also vars. *Aristolochia grandiflora, A. frimbriata* and *A. brasiliensis* etc.

ASPARAGUS
plumosa-

A relatively small climber, 1-2 m in length, leaves bristle-like in clusters, branches furnished with sharp thorns, flowers insignificant. Prefers a rich well-drained soil, semi-shaded, protected situation. Suitable for pots and hanging baskets.
Susceptible below 25°-32° F
Propagation: by seed or division
Saline tolerance: good
Water requirements: regular watering
Wind resistance: fair, requires protection from wind

ASPARAGUS
sprengeri-

Somewhat similar to *A. plumosa*, 1-1½ m in length, leaves filliform, branches furnished with thorns. Much coarser than *A. plumosa*.
Susceptible below 25°-32° F
Propagation: by seed or division
Saline tolerance: good
Water requirements: regular watering
Wind resistance: fair, requires protection

BOUGAINVILLIA
var-

Straggly, spiny, free-flowering, robust, scandent shrubs/climbers with alternate or opposite leaves. Flowers small, insignificant, bracts large, varying in colour. Main varieties, *B. glabra* and *B. spectabilis* with numerous hybrids. Prefers a rich, well-drained soil and a sunny protected situation. Should be cut-back regularly to obtain abundant flowers. Suitable for walls, trellises, trees, pillars, buildings and hedges.

Susceptible below 25°-32° F
Propagation: by cuttings
Saline tolerance: fair – variable
Water requirements: regular watering
Wind resistance: poor – fair, requires protection

CLERODENDRON *indica*-

Plate 23 (a)
A robust, quick-growing climber with small, dark green leaves and white flowers with long, projecting stamens. Hardy, and eminently suitable for hedges, windbreaks and screening, or for growing on walls. Requires regular maintenance as it tends to takeover and grow over other plants. Should not be planted too close to trees or shrubs as it is a coarse and voracious feeder.
Susceptible below 25° F
Propagation: by cuttings
Saline tolerance: excellent
Water requirements: regular watering during establishment, light watering after establishment
Wind resistance: good

CLITORIA *ternatea*-

Plate 23 (b)
A handsome, rather slender climber, 3-9 m in length, which tends to twine anti-clockwise. Leaves are usually scattered; flowers clear blue and sweet-pea-like in shape, which only last one day. Can exhibit chloritic symptoms in early stages. Suitable for growing over a wire netting or as a ground cover.
Susceptible below 25°-32° F
Propagation: by seed
Saline tolerance: reasonably good
Water requirements: regular watering during establishment, light watering after establishment
Wind resistance: reasonably good, requires protection in early stages

GLORIOSA *rothschildiana*-

Plate 23 (c)
A much-branched, twiner/climber, 1-3 m in length, with narrow leaves ending in a climbing tendril; flowers large, scarlet/yellow, very showy. Relatively slow growing. Requires a rich soil, an open and well-drained situation. Suitable for growing over supports.
Susceptible below 32° F
Propagation: by seed or division of horizontally growing rootstocks
Saline tolerance: fair
Water requirements: regular watering
Wind resistance: reasonably good, requires some protection

CLERODENDRON
indica (a)

CLITORIA
ternatea (b)

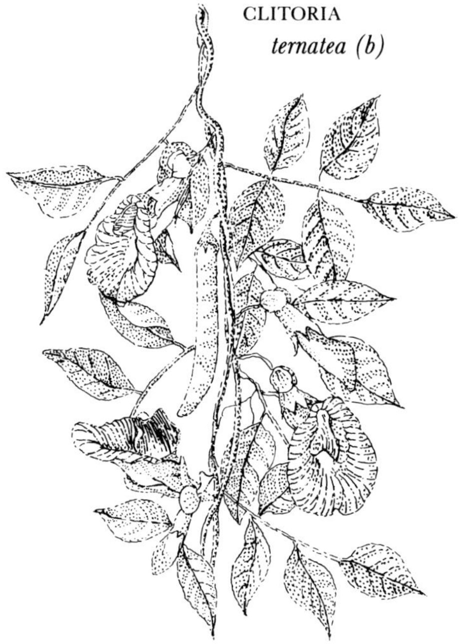

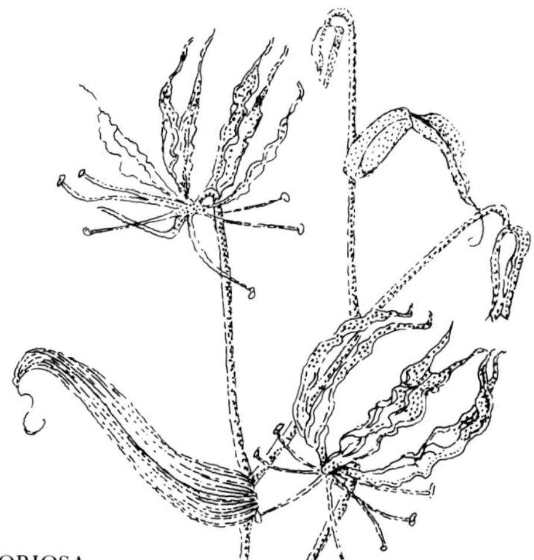

GLORIOSA
rothschildiana (c)

PLATE 23

IPOMOEA
var-

This genus includes both herbaceous and perennial species, numerous varieties, all of which should be planted where the flowers will receive morning sunlight, exception *I. alba* (Moon Flower).

IPOMOEA
horsfalliae-

Plate 24 (a)
A relatively large, woody perennial climber, 9-10½ m in length with palmate leaves and red, bell-shaped flowers produced in clusters. Tends to exhibit chloritic symptoms in early stages. Requires a rich, dry, well-drained soil and sunny situation. Suitable for training against houses, verandahs, trellis-work or wire netting.
Susceptible below 25°-32° F
Propagation: by seed or cuttings
Saline tolerance: fair
Water requirements: regular watering
Wind resistance: fair, requires some protection

IPOMOEA
learii-

Plate 24 (b)
A perennial climbing plant with 9-12 m long stems, small, evergreen, deeply cleft leaves and bright blue flowers. Hardy; prefers a well-drained, open situation. Suitable for training against a trellis or over wire netting.
Susceptible below 25°-32° F
Propagation: by seed or cuttings
Saline tolerance: fair
Water requirements: regular watering
Wind resistance: fair, requires some protection

IPOMOEA
pes caprea-

A robust, straggling vine with a tangled mass of wiry stems up to 18-24 m in length. Leaves large and leathery, resembling a goat's foot and funnel-shaped mauve/purple flowers. Very hardy, forms an excellent ground cover on banks and sandy shores.
Susceptible below 25°-32° F
Propagation: by cuttings
Saline tolerance: excellent
Water requirements: regular watering during establishment, light watering if and when required after establishment
Wind resistance: excellent

IPOMOEA
purpurea-

Plate 24 (c)
Annual climber with cordate-ovate, hairy leaves and large bell-shaped violet/purple flowers. Variation in colour – white, pink, red, dark blue, lilac, red-streaked and tri-coloured; also double and semi-double forms. Tends to exhibit chloritic symptoms in

IPOMOEA
learii (b)

IPOMOEA
horsfalliae (a)

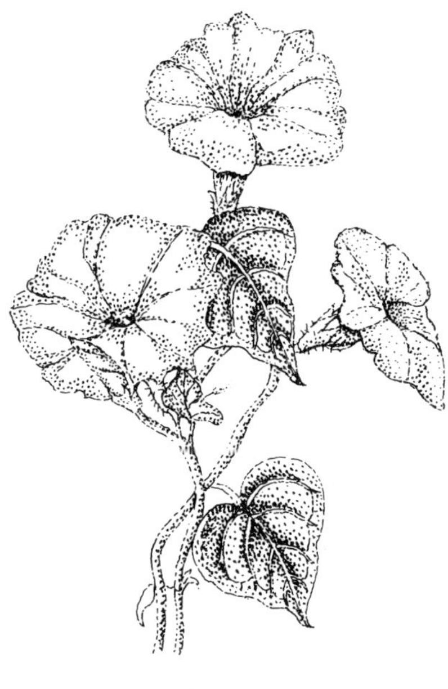

IPOMOEA
purpurea (c)

PLATE 24

early stages. Requires a rich, dry well-drained soil and sunny situation. Suitable for training against houses, verandahs, trellis-work or wire netting.
Susceptible below 25°-32° F
Propagation: by seed or cuttings
Saline tolerance: fair
Water requirements: regular watering
Wind resistance: fair, requires some protection

IPOMOEA *quamoclit var pennata*-

Plate 25 (a)
An attractive, dainty, relatively delicate, annual climber/twiner with long pinnate leaves and beautiful small, red, tubular flowers. Rather delicate under M. East conditions, prefers a rich, well-drained situation. Suitable for trellises or for growing on pillars.
Susceptible below 32° F
Propagation: by seed
Saline tolerance: relatively poor
Water requirements: regular watering
Wind resistance: poor, requires adequate protection from wind

JACQUEMONTIA *martii*-

Plate 25 (b)
A perennial climber with slender stems 4½-15 m in length. Leaves, small, dull green, cordate; flowers bell-shaped, light blue in colour, borne in clusters. Hardy, very free-flowering. Requires a dry, well-drained soil, sunny position and some protection. Suitable for training on walls, trellises, trees, pillars and buildings.
Susceptible below 25°-32° F
Propagation: by seed, cuttings or layers
Saline tolerance: reasonably good
Water requirements: regular watering
Wind resistance: poor, requires protection

JASMINUM *var*-

A large genus of over 300 species especially adapted to tropical and sub-tropical regions. Flowers vary in colour from white, yellow, to rose; white flowered types are usually very fragrant while yellow varieties are usually scentless. Generally speaking, most varieties are relatively hardy and free-flowering. They require a sunny or semi-shaded situation, a fairly rich soil and some protection from wind. Suitable for training on trellises, supports or on wire netting and trees.
Susceptible below 25°-32° F
Propagation: by seed, cuttings or layers
Saline tolerance: all varieties reasonably good

JASMINUM
officinale (c)

IPOMOEA
quamoclit var pennata (a)

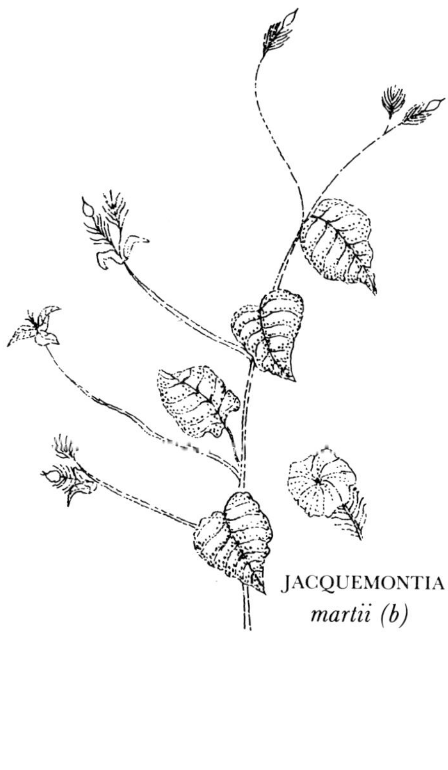

JACQUEMONTIA
martii (b)

PLATE 25

Water requirements: regular watering
Wind resistance: fair – all varieties require protection from wind

JASMINUM *officinale-*

Plate 25 (c)
White, 4½-15 m in length

JASMINUM *primulinum-*

White, 2-3 m in length

JASMINUM *pubescens-*

White, 2-3 m in length

JASMINUM *revolutum-*

Yellow, 2-4 m in length

JASMINUM *sambac-*

White, 1½-2 m in length

QUISQUALIS *indica-*

Plate 26 (a)
A heavy, quick-growing, semi-climber, which grows an erect shoot and then produces a basal climbing shoot; flowers borne in showy racemes. Flowers open white and change from pink to crimson; scented at night. Delicate under Middle East conditions, and tends to exhibit chloritic symptoms. Requires a rich and well-drained soil.
Susceptible below 40° F
Propagation: by cuttings
Saline tolerance: poor
Water requirements: regular watering
Wind resistance: poor, requires protection

SOLANUM *wendlandii-*

A robust, quick-growing climber, petioles and leaves prickly, leaves three-lobed or heart-shaped. Flowers lilac/blue borne in large panicles. The plant climbs by means of hooked prickles to a height of 4½-6 m. Rather delicate under Middle East conditions, tends to exhibit chloritic symptoms. Requires a fairly rich, well-drained open or semi-shaded situation.
Susceptible below 32° F
Propagation: by cuttings
Saline tolerance: poor
Water requirements: regular watering
Wind resistance: poor, requires protection

THUNBERGIA
grandiflora (c)

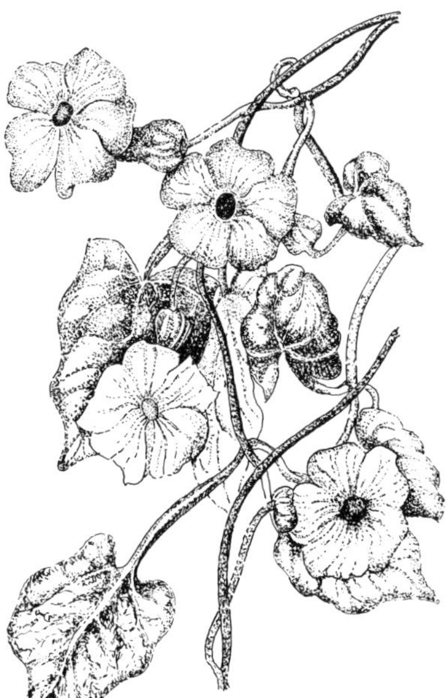

THUNBERGIA
alata (b)

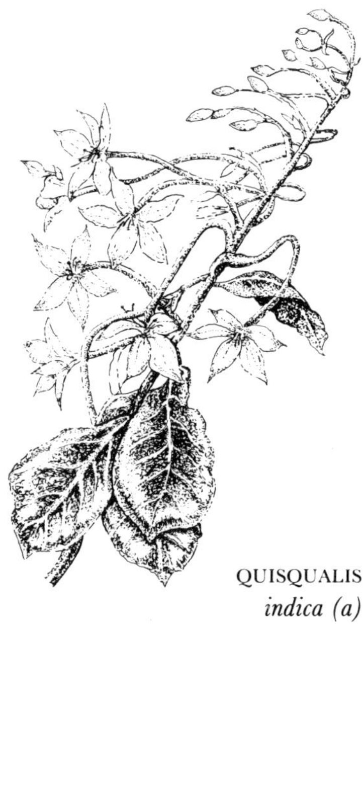

QUISQUALIS
indica (a)

PLATE 26

THUNBERGIA *alata*-

Plate 26 (b)

A perennial, herbaceous twiner 1-2 m in length, generally grown as an annual. Leaves ovate-cordate, flowers funnel-shaped, yellow or yellow/white with black spot in throat. Requires a rich, well-drained soil and an open situation. Suitable for growing on a trellis or wire netting.

Susceptible below 35° F
Propagation: by seed
Saline tolerance: reasonably good
Water requirements: regular watering
Wind resistance: fair, requires protection in early stages

THUNBERGIA *grandiflora*-

Plate 26 (c)

Robust, quick-growing, woody climber, 1-15 m in length. Leaves ovate to lanceolate, hairy; flowers sky blue or white, borne in festoons, free-flowering. Tends to be difficult under Middle East conditions and exhibits chloritic symptoms. Requires a dry, well-drained situation and suitable for growing on strong trellises against a wall or on supports.

Susceptible below 35° F
Propagation: by seed, cuttings or tubers
Saline tolerance: poor
Water requirements: regular watering
Wind resistance: poor to fair, requires protection

Note: For quick reference see Appendix C
For classification according to colour see Appendix C/1

Tuberous and herbaceous plants

AGAPANTHUS
africanus (a)

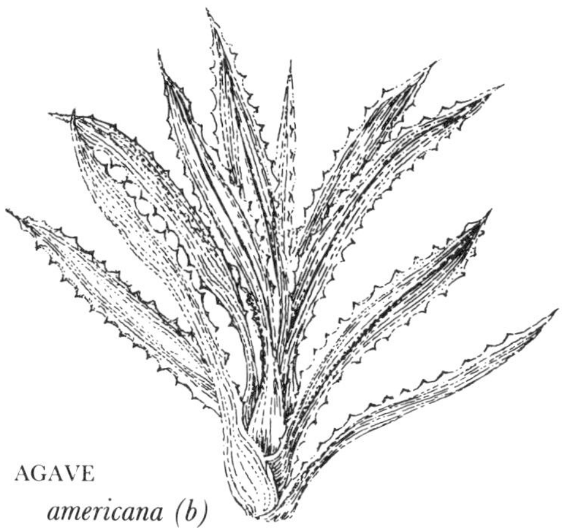

AGAVE
americana (b)

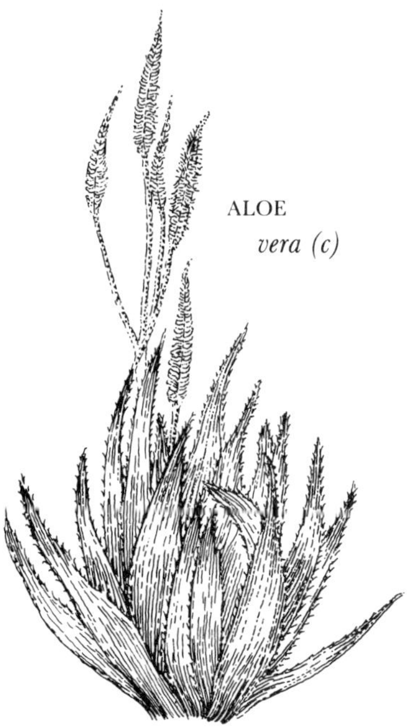

ALOE
vera (c)

PLATE 27

Latin name	Description
AGAPANTHUS *africanus*-	*Plate 27 (a)* Attractive, bulbous plant, 45-90 cm in height. Leaves long, narrow, 30-60 cm in length. Inflorescence umbellate with bunches of blue flowers on long stalks. Relatively delicate under M. East conditions, prefers a semi-shaded, well-drained situation. Suitable for grouping or planting in pots. Susceptible below 32° F Propagation: by division Saline tolerance: poor Water requirements: regular watering Wind resistance: poor
AGAVE *americana*-	*Plate 27 (b)* A robust, slow-growing, stemless plant, 1-2 m in height with a rosette of 20-60 tough, rigid, marginally prickly 1-1½ m long, green leaves with sharply pointed tips. Flowers yellowish/green on a terminal inflorescence 6-12 m in height. Var. *A. luteo-marginata* has broad leaves edged with yellow. Hardy, prefers open situation, suitable for specimen planting, bedding, border and boundary planting. Susceptible below 25° F Propagation: by offsets Saline tolerance: excellent Water requirements: regular watering during establishment Wind resistance: excellent
AGAVE *sisalana*-	Somewhat similar in growth to *A. americana*, with bluish/green leaves 1.5-1.8 m in length with sharp points. Hardy, prefers open situation. Suitable for specimen planting, grouping, border and boundary planting. When planted in depth, provides an impenetrable border and tends to attract snakes. Susceptible below 25° F Propagation: by offsets Saline tolerance: excellent Water requirements: regular watering during establishment Wind resistance: excellent
ALOE *var*-	*Plate 27 (c)* Handsome, succulent plants with an enormous variation in colour and size, somewhat similar in growth to the Agaves. Suitable for specimen planting, grouping or borders. Susceptible below 32° F Propagation: by seed or division

CANNA
lily (a)

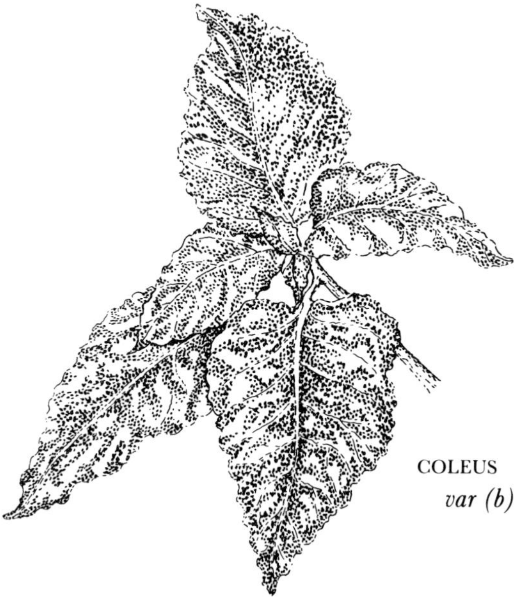

COLEUS
var (b)

CRINUM
var (c)

PLATE 28

Saline tolerance: reasonably good
Water requirements: regular watering during establishment
Wind resistance: excellent

CANNA *lily-*

Plate 28 (a)
Numerous hybrids, 1-2.5 m in height, leaves green or reddish/brown, flowers have a wide variation, white, pink, yellow, red and intermediate shades. Prefers an open, well-drained situation, a very rich soil and a high irrigation regime. Suitable for beds, borders and background planting.
Susceptible below 25°-32° F
Propagation: by seed or tubers
Saline tolerance: fair
Water requirements: regular, heavy watering
Wind resistance: poor

COLEUS *var-*

Plate 28 (b)
Decorative hybrids, leaves variegated. Suitable for group planting or growing in pots. Prefers a semi-shaded situation.
Susceptible below 32° F
Propagation: by seed
Saline tolerance: fairly poor
Water requirements: regular watering
Wind resistance: poor

CRINUM *var-*

Plate 28 (c)
Lilies with robust growth, long sword-shaped leaves and inflorescences of white or pink flowers. Requires a rich soil and semi-shaded situation. Suitable for potting.
Susceptible below 32° F
Propagation: by division
Saline tolerance: fair
Water requirements: regular watering
Wind resistance: poor

DIEFFENBACHIA *var-*

Plate 29 (a)
Ornamental foliage plants with variegated leaves. Prefers moist, shaded situation, suitable for potting.
Susceptible below 32° F
Propagation: by stem cuttings
Saline tolerance: relatively poor
Water requirements: regular watering
Wind resistance: poor

DIEFFENBACHIA
var (a)

EUCHARIS
amazonica (b)

GLORIOSA
var (c)

PLATE 29

EUCHARIS *amazonica-*

Plate 29 (b)

A bulbous plant with long-stalked leaves and handsome, faintly scented, white flowers. Prefers a shady situation: suitable for potting or bedding.

Susceptible below 32° F

Propagation: by bulbs

Saline tolerance: relatively poor

Water requirements: regular watering

Wind resistance: poor

GLORIOSA *var-*

Plate 29 (c)

Much-branched twiner/climbers 1-3 m in length, with narrow leaves ending in a climbing tendril: flowers large, scarlet/yellow, very showy. Relatively slow growing. Requires a rich soil and an open and well-drained situation. Suitable for growing over supports.

Susceptible below 32° F

Propagation: by seed or division of horizontally growing rootstocks

Saline tolerance: fair

Water requirements: regular watering

Wind resistance: reasonably good, requires some protection

HEMEROCALLIS *var-*

Plate 30 (a)

Robust, spreading perennials with narrow, linear leaves and attractive yellow or orange flowers. Prefers a rich soil, an open situation and a high irrigation regime. Suitable borders or beds.

Susceptible below 20°-25° F

Propagation: by division

Saline tolerance: fair

Water requirements: regular watering during growing season

Wind resistance: fair

HIPPEASTRUM *var-*

Plate 30 (b)

Bulbous plants with long, narrow leaves, flowers large, variable in colour and very attractive. Suitable for edging, border planting or potting.

Susceptible below 25° F

Propagation: by seed

Saline tolerance: fair

Water requirements: regular watering during growing season

Wind resistance: poor

HYLINE *gardneriana-*

Robust, free-flowering lilies with pure white slightly scented flowers. Suitable for borders or beds; prefer an open, well-drained situation.

HIPPEASTRUM
var (b)

HYMENOCALLIS
var (c)

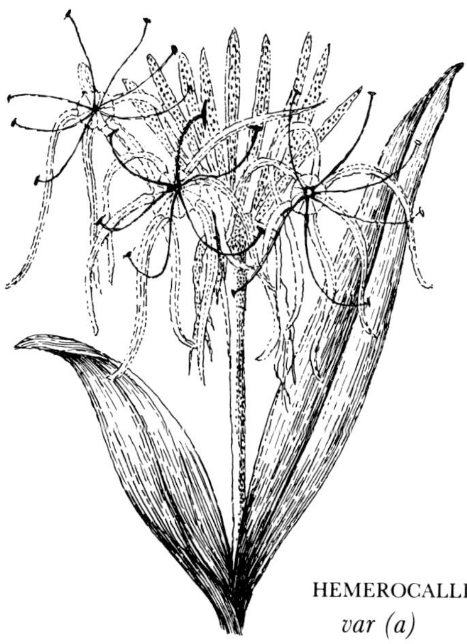

HEMEROCALLIS
var (a)

PLATE 30

Susceptible below 32° F
Propagation: by division or bulbs
Saline tolerance: excellent
Water requirements: regular watering during growing season
Wind resistance: good

HYMENOCALLIS
var-

Plate 30 (c)
Bulbous plants, 0.5-1.5 m in height with long leaves and white or yellowish/white, fragrant flowers borne in clusters on stout stalks. Prefers a moist, semi-shaded situation, some varieties delicate under M. East conditions. Suitable for bedding or borders.
Susceptible below 25° F
Propagation: by offsets
Saline tolerance: poor to fair
Water requirements: regular watering
Wind resistance: fair

KNIPHOFIA
var-

Plate 31 (a)
Attractive, grassy-leafed plants with conspicuous spikes of brilliant flowers. Flower colour varies between yellow, white, red and flame. Requires a rich and well-drained soil and an open situation. Suitable for bedding.
Susceptible below 25° F
Propagation: by seed or division
Saline tolerance: poor
Water requirements: regular watering during growing season
Wind resistance: poor

MIRABLIS
jalapa-

Plate 31 (b)
A herbaceous perennial often grown as an annual, 45-60 cm in height with a bushy habit and clusters of red, pink, white, yellow and streaked, scented flowers. The flowers open towards evening and close again in the morning. Hardy and suitable for beds and borders.
Susceptible below 40° F
Propagation: by seed
Saline tolerance: reasonably good
Water requirements: regular watering
Wind resistance: poor

OCIMUM
basilicum-

Hardy herb, 0.5-1 m in height; small leaves and purple flowers, prefers semi-shaded situation. Suitable in borders or planted along water channels.
Susceptible below 25° F
Propagation: by seed
Saline tolerance: reasonably good

KNIPHOFIA
var (a)

MIRABLIS
jalapa (b)

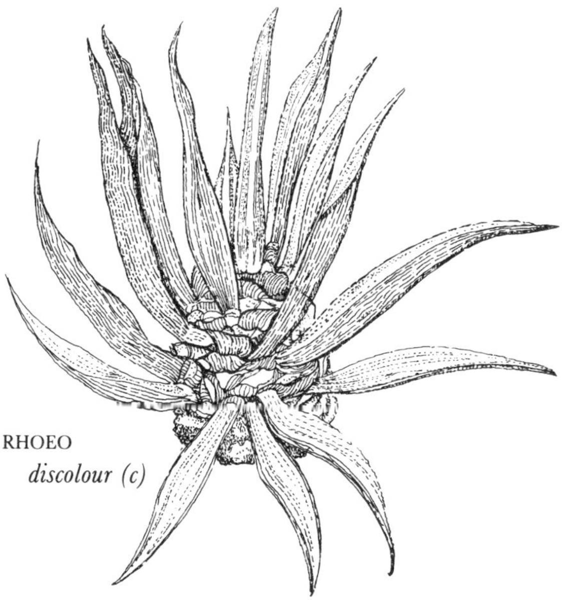

RHOEO
discolour (c)

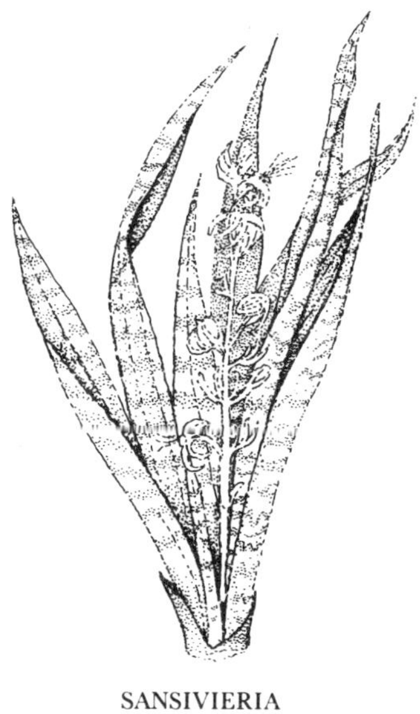

SANSIVIERIA
trifasciata (d)

PLATE 31

Water requirements: regular watering
Wind resistance: poor – fair, requires protection from wind

RHOEO *discolour-*

Plate 31 (c)

A small, perennial with narrow, lanceolate leaves which are bronze/green above and purplish/green/red below. Flowers greenish/white, insignificant: berries orange or red in colour. Prefers a rich, well-drained soil and open situation. Suitable for borders or pots.

Susceptible below 32° F
Propagation: by seed or division
Saline tolerance: good
Water requirements: regular watering
Wind resistance: good

SANSEVIERIA *trifasciata-*

Plate 31 (d)

Stiff, rather erect plants with long, linear-lanceolate, thick, leathery leaves. Hardy, prefers a dry, well-drained open situation. Suitable for pots.

Susceptible below 32° F
Propagation: by division or leaf cuttings
Saline tolerance: excellent
Water requirements: regular watering Feb-Oct, light watering after October if necessary
Wind resistance: excellent

VINCA *rosea-*

Plate 32 (a)

A hardy, herbaceous perennial, 30-75 cm in height with ovate shiny leaves and relatively large pink or white flowers. Hardy, prefers an open situation. Suitable for beds or borders.

Susceptible below 32° F
Propagation: by seed
Saline tolerance: good
Water requirements: regular watering
Wind resistance: fair

ZEBRINA *pendula-*

Plate 32 (b)

A creeping plant, 30-45 cm in length, with ovate-oblong, fleshy leaves, greyish/green above and purplish/red below. Flowers small, purplish/red in colour with white stamens.

Suitable for hanging baskets, pots or for edging.
Susceptible below 32° F
Propagation: by cuttings
Saline tolerance: fair
Water requirements: regular watering
Wind resistance: poor

VINCA
rosea (a)

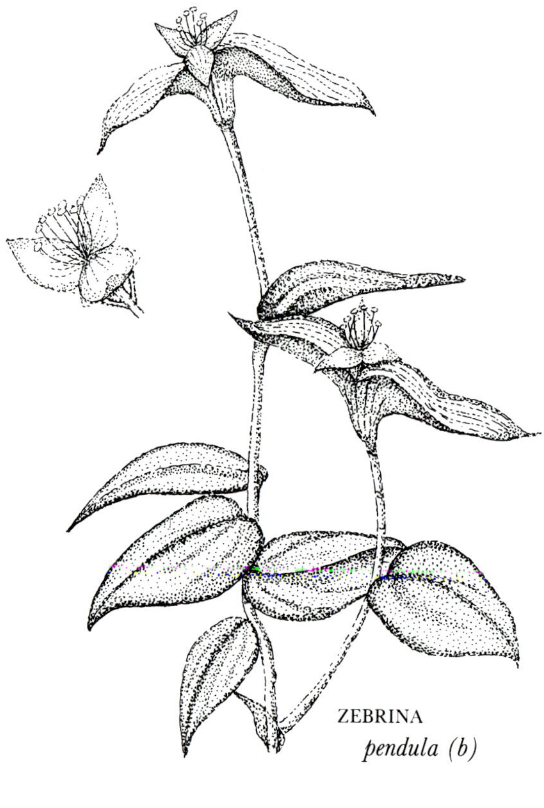

ZEBRINA
pendula (b)

PLATE 32

Herbs

Latin name	Description
ACHILLEA *santolina*-	A relatively low, flowering aromatic herb with yellow flowers and silvery pubescent leaves; approx. 10-15 cm in height, suitable for roadside planting or massing. Susceptibility: suitable desert conditions Propagation: by seed Saline tolerance: good Water requirements: light irrigation during establishment, does well in areas with rainfall of 100-150 mm Wind resistance: good
ANETHUM *graveolens*-	A herb long cultivated in Saudi Arabian gardens, 60 cm high with umbels of yellow flowers in November. Used for flavouring and dates. Ornamental in masses or as a specimen. Susceptibility: suitable desert conditions Propagation: by seed or division Saline tolerance: excellent Water requirements: arid to semi-arid, light irrigation during establishment Wind resistance: needs some shelter, subject to wind-pruning in open fields
CITRULLUS *colocynthis*-	An annual herb with relatively large green/white striped fruit that turn bright yellow on ripening. *C. colocynthis* is well known for the bitterness of its fruits which are often used medicinally. Not very suitable for desert landscaping, may serve for background planting on dry bare areas. Susceptibility: suitable desert conditions Propagation: by seed Saline tolerance: excellent Water requirements: light irrigation during establishment Wind resistance: excellent
HERNIARIA *hemistemon*-	A perennial herb, found in sandy, especially coastal, areas. Has many tomentose branches and minute flowers; the small leaves are also sometimes hairy on the underside and brown. It is found in gypsified desert soil. Susceptibility: suitable desert conditions Propagation: by seed or division Saline tolerance: excellent Water requirements: arid conditions Wind resistance: excellent

OCIMUM
basilicum- A hardy herb, ½-1 m in height; small leaves and purple flowers. Prefers a semi-shaded situation. Suitable in borders or along water channels.
Susceptibility: susceptible below 25° F
Propagation: by seed
Saline tolerance: reasonably good
Water requirements: regular watering
Wind resistance: fair – requires good protection

PLANTAGO
ovata- Annual herb, usually found growing in sandy areas. The seeds, when placed in water release a mucilage and are used medicinally. Could be used to break up small, bare, sandy patches of ground.
Susceptibility: suitable desert conditions
Propagation: by seed
Saline tolerance: excellent
Water requirements: desert conditions
Wind resistance: excellent

PTERANTHUS
dichtomus- A semi-dwarf flowering herb with fleshy stems and insignificant flowers. Provided seed is available, could be used in conjunction with other desert flora in a natural landscape.
Susceptibility: suitable desert conditions
Propagation: by seed
Saline tolerance: good
Water requirements: desert conditions
Wind resistance: excellent

ROBBAIREA
prostrata- A low perennial herb found in hammadas and sandy areas. (Syn. *Robbairea delileana)* up to 30 cm high. Has many branches and whiteish pink flowers in January-April. The seeds are smooth and shiny and the woody stems often prostrate. Can be used in desert landscapes with supplemental water.
Susceptibility: suitable desert conditions
Propagation: seed
Saline tolerance: excellent
Water requirements: desert conditions with some supplement to rainfall as run-off
Wind resistance: very good

RUMEX
cyprius- An annual herb with relatively fleshy leaves, occasionally seen growing on hammadas and in the foothills, which could possibly be used to break up small patches of bare soil.

Susceptibility: suitable desert conditions
Propagation: seed
Saline tolerance: good
Water requirements: light irrigation during initial establishment
Wind resistance: excellent

URTICA
urens-

Annual herb found in fog belts. The flowers are inconspicuous, and it has stinging hair and leaves. The foliage is used for fodder and also has some medicinal use.
Susceptibility: suitable desert conditions
Propagation: seed
Saline tolerance: excellent
Water requirements: probably needs some supplemental water
Wind resistance: good

Indigenous grasses with low water requirements

Latin name	Description
AEGILOPS *kotschyi-*	An indigenous, prostrate annual usually found growing on hamadas and sandy plains. Provided seed is obtainable from the scattered mosaic, goat grass may prove to be a reasonably good substitute for covering banks and roadside planting. Susceptibility: suitable desert conditions Propagation: by seed Saline tolerance: excellent Water requirements: light watering during establishment Wind resistance: excellent
ARISTIDA spp:	
(a) ARISTIDA *coerulescens-*	Indigenous annual or short-lived perennial, rhizomatic, bunch grass; excellent sandbinder, usually found growing on sand dunes, low escarpments or in the foothills. Susceptibility: suitable desert conditions Propagation: sprigs or division Saline tolerance: excellent Water requirements: minimal; desert conditions Wind resistance: excellent
(b) ARISTIDA *pungens-*	A slow-growing, prostrate grass sometimes used for fodder. Can be used as a soil-binder on banks. Seeds may be difficult to obtain locally. Susceptibility: suitable desert conditions Propagation: by seed, sprigs or division Saline tolerance: excellent Water requirements: minimal, desert conditions Wind resistance: excellent
(c) ARISTIDA *obtusa-*	An indigenous, prostrate perennial, densely tufted, a tough, palatable, drought resistant grass found growing on sand dunes, stream beds and in the foothills. Suitable as a sand-binder to reduce wind erosion. Susceptibility: suitable desert conditions Propagation: sprigs or division Saline tolerance: excellent Water requirements: minimal, desert conditions Wind resistance: excellent

CENCHURUS
ciliaris-

A relatively short, indigenous perennial, densely tufted, both rhizomatic and stoloniferous; drought resistant and persistant. Forms an excellent sand-binder, usually found growing in stony or sandy soils.
Susceptibility: suitable desert conditions
Propagation: seed or sprigs
Saline tolerance: excellent
Water requirements: light irrigation
Wind resistance: excellent

Note:
Seeds of the short Western-Australian type of *Cenchurus ciliaris* may do well under local conditions.

CYMBOPOGON
schoenanthus-

An indigenous perennial, strongly rooted bunch grass usually found growing on sand dunes, escarpments and in wadis under hostile conditions. Forms an excellent sand-binder.
Susceptibility: suitable desert conditions
Propagation: seed, if obtainable
Saline tolerance: excellent
Water requirements: light irrigation during establishment: eminently suitable in areas with 150 mm rainfall
Wind resistance: excellent

CYNODON
dactylon-

Cynodon dactylon (Bermuda Grass), with its wide range of varieties is in all probability the most important species of grass for turf production in the M. East. This long-lived perennial, with its spreading habit of growth, is very adaptable to a wide range of soils and climatic conditions and is widely distributed throughout the M. East in areas where there is sufficient precipitation.
Bermuda Grass is a "warm season" grass, and although heat and drought hardiness is excellent, low temperature tolerance is poor and it tends to lose its colour and become dormant when the soil temperature drops below 10° C. Provided adequate irrigation is available and a high standard of maintenance is practiced, Bermuda Grass can be used for fodder production, and under favourable conditions yields of 36 tons per ha. per annum, with a crude protein content varying between 11.00-17.00%, are not unknown.
Susceptibility: loss of colour and dormancy when soil temperature falls below 10° C
Propagation: runners, stolons, seed, dibbling and sodding

Saline tolerance: excellent
Water requirements: regular irrigation
Wind resistance: excellent

DACTYLOCTENIUM
aegyptiacum-

An indigenous, low-growing stoloniferous grass usually found growing in wadis and damp places. Can form a relatively dense sward under irrigation, shade tolerant and drought resistant. Suitable for stabilisation of banks or open places.
Susceptibility: suitable desert conditions
Propagation: sprigs or division
Saline tolerance: good
Water requirements: regular irrigation
Wind resistance: excellent

DANTHONIA
forskali-

An indigenous, deep-rooted, drought resistant, perennial grass usually found growing on sandy plains and on escarpments. Suitable for inclusion in desert landscapes.
Susceptibility: suitable desert conditions
Propagation: seed may be difficult to obtain locally and a suitable strain from Australia may do well under local conditions
Saline tolerance: good
Water requirements: minimal; will provide a reasonable sward under irrigation and will also grow under desert conditions where the rainfall is above 150 mm
Wind resistance: good

ERAGROSTIS spp:

An indigenous perennial, rhizomatic grass with tough, stiff stems and leaves which form a fairly good cover under suitable conditions.

(a) ERAGROSTIS
abyssinica-

A relatively tall grass that is cultivated for grain. Not really suitable for desert landscapes; could possibly be used for background planting under light irrigation.
Susceptibility: suitable under desert conditions
Propagation: seed
Saline tolerance: fair
Water requirements: light, regular irrigation, approx. 150-200 mm over season
Wind resistance: fair

(b) ERAGROSTIS
binnata-

An indicator plant that grows in alkaline areas and poorly drained pans. Suitable as a soil binder.
Susceptibility: suitable under desert conditions
Propagation: seed, stolons or rhizomes
Saline tolerance: excellent
Water requirements: light irrigation during establishment; should do well in an area with rainfall of 200 mm
Wind resistance: good

LASIRUS
hirsutus-

An indigenous perennial, rhizomatic grass usually found growing on escarpments or in the foothills. Spreads relatively slowly and can be used as a sand binder.
Susceptibility: suitable under desert conditions
Propagation: runners or sprigs
Saline tolerance: excellent
Water requirements: minimal irrigation during establishment
Wind resistance: excellent

PANICUM
turgidum-

An indigenous, low, zerophytic, tussock-type grass which spreads by rhizomes and is often found growing in rocky, hilly and sandy areas, in wadis and hammadas. Suitable for roadside planting as it forms a reasonable cover under suitable conditions.
Susceptibility: suitable under desert conditions
Propagation: by seed, sprigs, division or rhizomes
Saline tolerance: reasonably good
Water requirements: light irrigation during establishment. Does well in areas with a rainfall of 100 mm
Wind resistance: reasonably good

SCHISMUS spp:

(a) SCHISMUS
arabicus-

An indigenous, relatively low-growing, rather rare grass usually found in sandy locations in S. Arabia. Can be used in desert landscapes providing seed or planting materials are available.
Susceptibility: suitable under desert conditions
Propagation: seed or division
Saline tolerance: good
Water requirements: desert conditions
Wind resistance: good

(b) SCHISMUS
barbatus- Somewhat similar to *S. arabicus,* found in loose, sandy habitats more noticeable in N. Africa than in Saudi Arabia.
Susceptibility: suitable under desert conditions
Propagation: seed or division
Saline tolerance: good
Water requirements: desert conditions
Wind resistance: good

STIPA
tortilis- An indigenous, perennial grass 40-90 cm in height with prominent needle-like bristles, which is used for fodder locally. Suitable for background or cover planting on compact sands or gravels.
Susceptibility: suitable under desert conditions
Propagation: seed
Saline tolerance: fair
Water requirements: minimal during establishment
Wind resistance: fair

Indigenous plants with low water requirements

Latin name | Description

ADIANTUM
capillus-veneris-

Hardy trailing fern with pinnate fronds 15-30 cm in length occasionally found growing along the edge of water channels or in the masonry of walls. Suitable for pots or hanging baskets.
Susceptible below 40° F
Propagation: rhizomes
Saline tolerance: poor
Water requirements: regular irrigation
Wind resistance: poor

(a) AGAVE
americana-

A robust, slow-growing, stemless plant, 1-2 m in height, with a rosette of 20-60 tough, rigid, marginally prickly 1-1½ m long, green leaves with sharply pointed tips. Flowers yellowish/green on a terminal inflorescence 6-12 m in height. Var. *A. luteo-marginata* has broad leaves edged with yellow. Hardy, prefers open situation, suitable for specimen planting, bedding, border and boundary planting.
Susceptible below 25° F
Propagation: by offsets
Saline tolerance: excellent
Water requirements: regular watering during establishment
Wind resistance: excellent

(b) AGAVE
sisalana-

Somewhat similar in growth to *A. americana*, with bluish/green leaves 1.5-1.8 m in length with sharp points. Hardy, prefers open situation. Suitable for specimen planting, grouping, border and boundary planting. When planted in depth, provides an impenetrable border and tends to attract snakes.
Susceptible below 25° F
Propagation: by offsets
Saline tolerance: excellent
Water requirements: regular watering during establishment
Wind resistance: excellent

ALLIUM spp:

(a) ALLIUM
flavum-

Wild onion, with fine leaves, approx. 15-20 cm in height, with yellow flowers occasionally found at higher altitudes in Saudi Arabia. Can be used with grass to provide a natural effect.
Susceptibility: suitable at higher altitudes
Propagation: bulbs

Saline tolerance: fair
Water requirements: regular irrigation
Wind resistance: good

(b) ALLIUM
momtamea- A wild, semi-dwarf onion with fine leaves approx. 10-15 cm in height. Flowers spherical, mauve/purple in colour.
Susceptibility: suitable at higher altitudes, Western Highlands, Hail etc.
Saline tolerance: fair
Water requirements: regular irrigation
Wind resistance: good

ASTERICUS
pygmaeus- A small, perennial plant with daisy-like flowers and grey leaves which usually appear during the winter rains. Suitable for massing provided seeds are obtainable.
Susceptibility: suitable desert conditions
Propagation: seed
Saline tolerance: fair
Water requirements: regular irrigation during establishment
Wind resistance: fair to good

ATRACTYLIS
serratuloides- A thorny annual with a small, insignificant flower which could be used in desert landscaping operations to cover open sandy areas provided seed is available.
Susceptibility: suitable desert conditions
Propagation: seed
Saline tolerance: good
Water requirements: desert conditions
Wind resistance: excellent

CACTUS- Succulents, fleshy plants with extensive root systems and usually armed with thorns. Eminently suitable for concentration for effect or interplanting with other succulents on dry, unirrigated areas.
Susceptibility: varies with type, generally speaking, cacti can withstand a wide variation of temperatures
Propagation: seed, cuttings, grafts
Saline tolerance: good to excellent
Water requirements: light irrigation during establishment and occasional irrigation if and when necessary after establishment
Wind resistance: excellent

CARRICHTERA
annua- Dwarf, grass-like annual with cream coloured flowers which is often found growing in the foothills or on hammadas.
Susceptibility: suitable desert conditions
Propagation: by seed or transplanting
Saline tolerance: fair
Water requirements: light, regular watering during establishment
Wind resistance: excellent

Note:
Extremely doubtful if seed or planting material will be readily available.

CISTANCHE
lutea- An annual hyacinth with scented mauve and yellow flowers which could be used for effect on dry areas.
Susceptibility: suitable desert conditions
Propagation: bulbs
Saline tolerance: fair
Water requirements: light irrigation during growing period
Wind resistance: excellent

CYCAS
revoluta- A distinct species of the Cycadaceae family, more suitable for establishment in the sub-tropics. The long, blue/green pinnate leaves grow out of a thick base, and growth in the Middle East is exceptionally slow. Suitable as specimen plants.
Susceptible below 40° F
Propagation: by seed or suckers
Saline tolerance: fair
Water requirements: regular irrigation
Wind resistance: fair

CYPERUS spp:

(a) CYPERUS
conglomeratus- A perennial, grass-like sedge often found growing on sand dunes and sabkhahs. Unsuitable for desert landscaping, can be used to stabilise sandy slopes or on sabkhakas to break up the cemented surface.
Susceptibility: suitable desert conditions
Propagation: runners or rhizomes
Saline tolerance: excellent indicator plant as it also grows well on areas with high saline watertable
Water requirements: desert conditions
Wind resistance: excellent

(b) CYPERUS

laevigatus- A perennial, grass-like sedge often found growing in wadis, and consists of a number of stalks with needle-like leaves.
Remarks as for *conglomeratus*.

DIPLOTAXIS

harra- A dwarf annual often found growing on sand dunes and hammadas. Provided seed or plants are available, may prove to be a suitable sand binder.
Susceptibility: suitable desert conditions
Propagation: seed, if available
Saline tolerance: excellent
Water requirements: desert conditions
Wind resistance: excellent

EREMOBIUM

aegyptiacum: v lineare- An annual, approx. 30 cm in height, with fleshy leaves and mauve/violet flowers, usually found growing in sandy areas. Will in all probability be difficult to establish as seed and planting material will be difficult to obtain.
Susceptibility: suitable desert conditions
Propagation: seed or plants
Saline tolerance: excellent
Water requirements: desert conditions
Wind resistance: excellent

FIMBRISTYLIS spp- A relatively tall reed, with sharp, pointed leaves usually found growing on irrigation channels. Could be used as a windbreak where water is not a problem.
Susceptibility: suitable desert conditions
Propagation: separation
Saline tolerance: fair
Water requirements: heavy, continuous irrigation
Wind resistance: excellent

GYMNARRHENA

micrantha- A low, flowering annual often found growing in the foothills. Suitable for establishment on dry, sandy areas providing seed is obtainable.
Susceptibility: suitable desert conditions
Propagation: seed
Saline tolerance: good
Water requirements: light irrigation during establishment
Wind resistance: excellent

LEYSSERA
capillifolia- Dwarf, flowering annual: flowers relatively insignificant. Provided seed is obtainable, could be used on bare, sandy areas.
Susceptibility: suitable desert conditions
Propagation: seed
Saline tolerance: good
Water requirements: desert conditions
Wind resistance: excellent

MESEMBRYANTHEMUM spp:

(a) MESEMBRYANTHEMUM
crystallinum- This annual variety, which is generally known as 'Gasoul' in Saudi Arabia, has glass-like papillae on the leaves. The flowers are creamy-white and the edible fruit, which have a narcotic content, are chewed or used as a stock feed. Suitable for establishment on dry, sandy areas.
Susceptible below 32° F
Propagation: seed or slips
Saline tolerance: good
Water requirements: light irrigation during establishment
Wind resistance: excellent

(b) MESEMBRYANTHEMUM
forskali- Somewhat similar to *M. crystallinum:* fruits are used by the Bedouins. Suitable for establishment on dry, sandy areas.
Susceptible below 32° F
Propagation: seed or slips
Saline tolerance: good
Water requirements: light irrigation during establishment
Wind resistance: excellent

MONSONIA
nivea- A dwarf annual with white/rose/purple flowers and long, beaked fruits suitable for establishment on bare, sandy areas.
Susceptibility: suitable desert conditions
Propagation: seed
Saline tolerance: good
Water requirements: desert conditions
Wind resistance: good

MIRABLIS
jalapa- A herbaceous perennial, often grown as an annual; 45-60 cm in height with a bushy habit and clusters of red, pink, white, yellow and streaked scented flowers. The flowers open towards evening

and close again in the morning. Hardy and suitable for beds and borders.
Susceptible below 40° F
Propagation: seed
Saline tolerance: reasonably good
Water requirements: regular watering
Wind resistance: poor

OPUNTIA spp-

A very wide variety of succulents that vary considerably in size and habit, and are usually characterised by much branched, swollen stems usually covered with sharp, very prickly spines. The water-filled stems and the structure of these succulents tends to reduce transpiration and makes these plants drought resistant, and enables them to survive in hot, arid areas. The root system of varieties that do well under desert conditions are usually shallow and relatively extensive. Flowers are usually bright and attractive, and the fruit of some opuntias is edible. *O. ficus indica*, which is spineless, provides fodder, and thorny varieties can also be used for fodder provided they are despined. Opuntias are eminently suitable as barrier hedges and soil stabilisation on slopes.
Susceptibility: suitable for desert conditions, should not be grown at altitudes above 700-800 m
Propagation: cuttings
Saline tolerance: excellent
Water requirements: light irrigation during establishment, does well in areas with 100-150 mm rainfall
Wind resistance: excellent

ORCHIS *palustris*-

A hardy, ground orchid with mauve flowers, often found growing in marshy areas. This orchid is of little or no use in landscaping operations unless it is used together with other water-loving plants on open water surrounds.
Susceptibility: desert conditions
Propagation: plants
Saline tolerance: good
Water requirements: continuous supply of water
Wind resistance: poor

PHILIPAEA *lutea*-

An annual parasitic succulent with a thick, yellowish-green flower spike which is often found growing in soft sandy areas. Not exactly ornamental or suitable for desert landscaping, however it could be used to break up small, bare, sandy areas

Susceptibility: suitable desert conditions
Propagation: rootlets or underground stems
Saline tolerance: good
Water requirements: desert conditions
Wind resistance: good

PHRAGMITES
communis-

A tall, tough reed 150-180 cm in height with an extensive root system which grows along water channels, in shallow water or swamps. If grown on water channels or the sides of wadis, it can form a very effective windbreak.
Susceptibility: suitable desert conditions
Propagation: rhizomes or by division
Saline tolerance: excellent, will grow under exceptionally saline conditions
Water requirements: high irrigation regime or continuous supply
Wind resistance: excellent

PITHURANTHOS
tortuosus-

Dwarf, grass-like annual that does well in sandy areas and could provide light ground cover on dry, sandy areas.
Susceptibility: suitable desert conditions
Propagation: seed or sprigs
Saline tolerance: excellent
Water requirements: light irrigation during initial establishment
Wind resistance: excellent

SCABIOSA
aucheri-

An annual scabious often found growing on hammadas and in the foothills; could be included in desert landscapes.
Susceptibility: suitable desert conditions
Propagation: by division
Saline tolerance: good
Water requirements: light irrigation during initial establishment
Wind resistance: excellent

SCORZONERA
judaica-

A semi-dwarf, tuberous rooted annual usually seen growing in the foothills. The flowers are yellow and the roots are edible, and this plant could be considered for inclusion in desert landscapes.
Susceptibility: suitable desert conditions
Propagation: seed
Saline tolerance: good
Water requirements: light irrigation in initial establishment
Wind resistance: excellent

TETRADICLIS
salsa-

A low-growing annual normally found growing in exceptionally saline soils, could be used to break up sandy, saline areas.
Susceptibility: suitable desert conditions
Propagation: seed
Saline tolerance: excellent
Water requirements: desert conditions
Wind resistance: excellent

Appendices

Key:-

H	Hardy, suitable for Middle East
FH	Fairly hardy, suitable for Middle East, may require protection in early stages
T	Tolerant, saline soils, water and desert conditions
FT	Fairly tolerant, saline soils, water and desert conditions
O	Ornamental
D	Delicate
Var	Varieties

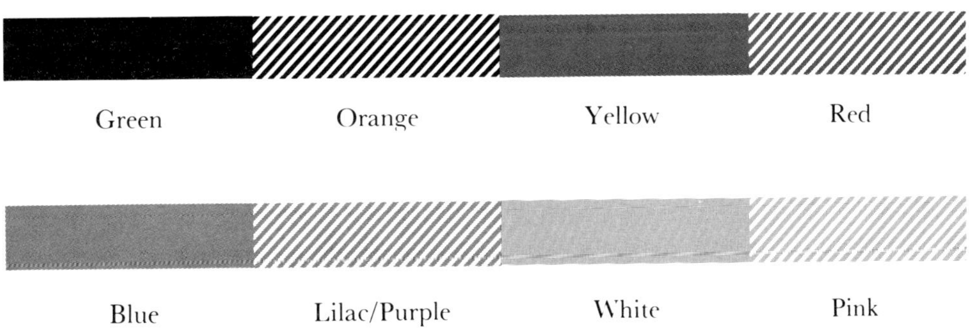

Flowering and foliage trees suitable for the Middle East

LATIN NAME	KEY	PROPAGATION
ACACIA *arabica*	H. T. O.	Seed
ACACIA *cyanophylla*	FH. FT. O.	Seed or cuttings
ACACIA *decurrens dealbata*	FH. FT. O.	Seed or cuttings
ACACIA *decurrens mollis*	H. T. O.	Seed or cuttings
ALBIZZIA *lebbec*	H. T. O.	Seed or cuttings
AZADIRACHTA *indica*	H. T. O.	Seed
BAUHINIA *alba var candida*	FH. FT. O.	Seed
BAUHINIA *purpurea*	FH. FT. O.	Seed
BAUHINIA *punctata*	FH. FT. O.	Seed
BAUHINIA *tomentosa*	FH. FT. O.	Seed
BUTEA *frondosa*	H. FT. O.	Seed
CALLISTEMON *lanceolatus*	H. T. O.	Seed
CASSIA *fistula*	D. FT. O.	Seed
CASSIA *grandis*	D. FT. O.	Seed
CASSIA *multijuga*	FH. FT. O.	Seed or cuttings
CASUARINA *cunninghamiana*	H. T. O.	Seed

APPENDIX A

REMARKS

Suitable desert conditions, specimen or background planting

Suitable for background planting in a partially protected location

Suitable for grouping or background planting in partially protected location

Suitable for specimen or background planting

Suitable large gardens and parks, branches brittle, should not be planted on avenues or near houses

Suitable avenues, desert conditions, or background planting

Suitable for massing or background planting in protected location

Suitable for massing or background planting in protected location

Suitable for massing or background planting in protected location

Suitable for massing or background planting in protected location

Suitable for background planting

Suitable for specimen, inner-line planting or grouping

Suitable for house gardens and parks in protected environment

Suitable for gardens in well-drained areas in a protected location

Suitable for small gardens in a protected environment

Suitable desert conditions, screening, windbreaks

LATIN NAME	KEY	PROPAGATION
CASUARINA *equisetifolia*	H. T.	Seed
CERATONIA *siliqua*	H. T.	Seed
CORDIA *abyssinica*	FH. FT. O.	Seed
DALBERGIA *sissoo*	H. FT. O.	Seed
DELONIX *regia*	H. T. O.	Seed, cuttings, large branches
ERYTHRINA *indica*	H. T. O.	Seed, cuttings or large branches
ERYTHRINA *var*	H. T. O.	Seed, cuttings or large branches
EUCALYPTUS *citriodora*	H. T. O.	Seed
EUCALYPTUS *var*	H. T. O.	Seed
FICUS *benghalensis*	H. T. O.	Cuttings
FICUS *benjamina*	H. FT. O.	Cuttings
FICUS *comosa*	H. FT. O.	Cuttings
FICUS *religiosa*	H. T. O.	Cuttings
GLIRICIDIA *maculata*	FH. T. O.	Seed or cuttings
JACARANDA *mimosaefolia*	D. O.	Seed
LEUCAENA *glauca*	FH. FT.	Seed
MELIA *azedarach*	H. T. O.	Seed or cuttings
MILLINGTONIA *hortensis*	H. T. O.	Cuttings
MORINGA *pterygosperma*	FH. T. O.	Seed
PARKINSONIA *aculeata*	FH. T. O.	Seed
PHOENIX *dactylifera*	H. T. O.	Suckers or offshoots
PISTACIA *atlantica*	H. FT. O.	Seed or cuttings

REMARKS

Suitable desert conditions, screening, windbreaks and coastal protection

Suitable desert conditions, slow growing, growth rate improves in protected environment

Suitable for massing, background or specimen planting

Suitable desert conditions, large gardens, avenues, parks

Suitable desert conditions, susceptible to frost

Suitable for background planting, rapid growth in protected location

Suitable for background planting, rapid growth in protected location

Suitable for avenues, roadside planting, grouping, parks

Suitable for avenues, roadside planting, grouping, parks

Suitable for large gardens, parks and roadside planting

Suitable for specimen or avenue planting or for shade in large gardens

Suitable for shade in large gardens

Suitable for specimen or background planting in large gardens or parks

Suitable for screening, background planting

Suitable for grouping, avenue or roadside planting in protected environment

Suitable for screening or background planting

Suitable for avenue planting, inner line and background planting

Suitable for avenues or specimen planting in protected environment

Suitable for specimen or background planting, screening and light shade

Suitable for avenue planting, inner line and background planting

Suitable for avenue or background, or planting in depth

Suitable as specimen, background and planting on berms

LATIN NAME	KEY	PROPAGATION
PITHECOLOBIUM *dulce*	H. T. O.	Seed
PLUMERIA *acutifolia*	FH. FT. O.	Cuttings
PLUMERIA *rubra*	FH. FT. O.	Cuttings
PONGAMIA *glabra*	FH. FT. O.	Seed
POPULUS *alba*	FH. FT.	Seed or cuttings
POPULUS *euphratica*	H. FT.	Seed or cuttings
PROSOPIS *juliflora*	H. T.	Seed
SALIX *babylonica*	FH. FT. O.	Cuttings
SCHINUS *molle*	H. T. O.	Seed or cuttings
SCHINUS *terebinthifolius*	H. T. O.	Seed or cuttings
SESBANIA *grandiflora*	H. T. O.	Seed
SPATHODEA *campanulata*	FH. FT. O.	Seed, suckers or cuttings
SPATHODEA *nilocta*	FH. FT. O.	Seed
TAMARDINDUS *indica*	H. T. O.	Seed
TAMARIX *var*	H. T. O.	Cuttings
TERMINALIA *catappa*	H. T. O.	Seed
THESPESIA *populnea*	H. T. O.	Seed
THEVETIA *nerefolia*	H. FT. O.	Seed or cuttings
TIPUANA *tipu*	H. T. O.	Seed
ZIZYPHUS *spina christi*	H. T.	Seed or cuttings
ZIZYPHUS *spina vulgare*	H. T.	Seed

REMARKS

Suitable for large house gardens, roadside planting or planting in close proximity to the seashore; tolerates relatively high watertables

Suitable for parkway, roadside and background planting, especially in coastal areas

Suitable for specimen grouping, background planting or screening

Suitable for street planting and light shade

Suitable for screening, windbreaks and boundary planting

Suitable for screening, windbreaks and boundary planting

Suitable for shade, avenue or boundary planting

Suitable for beds, borders, grouping or specimen planting in protected environment

Suitable for borders, grouping or specimen planting in protected environment

Suitable for street planting, background or massing in depth; tends to be rather delicate in saline areas or areas irrigated with poor quality water

Suitable avenue planting, windbreaks and screening in moist situations

Suitable avenue planting, windbreaks and screening in moist situations

Suitable for screening and windbreak planting in depth

Suitable for specimen planting ins protected environment

Suitable for screening, background or specimen planting

Suitable for background planting or screening

Suitable for specimen, background planting and screening, especially in coastal areas

Suitable for large gardens and parks, unsuitable in areas with high watertables

Suitable for roadside or background planting: unsuitable in areas with a high watertable

Suitable for roadside or background, in large gardens or parks

Suitable for screening or planting in sandy, arid areas or in areas where strong, saline-laden winds are experienced

Classification of trees according to colour

ALBIZZIA *lebbec*
CASUARINA *cunninghamiana*
CASUARINA *equisetifolia*
CERATONIA *siliqua*
DALBERGIA *sisoo*
EUCALYPTUS *var citriodora*
FICUS *benghalensis*
FISCUS *benjamina*
FICUS *comosa*
FICUS *religiosa*
LEUCANIA *glauca*
PHOENIX *dactylifera*
PISTACHIA *atlantica*
PITHECOLOBIUM *dulce*
POPULUS *alba*

POPULUS *euphratica*
PROSOPIS *juliflora*
SALIX *babylonica*
SCHINUS *molle*
SCHINUS *terebinthifolius*
TAMARINDUS *indica*
TERMINALIA *catappa*
ZIZYPHUS *spina-christy*
ZIZYPHUS *spina-vulgare*

BUTEA *frondosa*
SPATHODEA *campanulata*
SPATHODEA *nilocta*
THEVETIA *nerefolia*

ACACIA *arabica*
ACACIA *cyanophylla*
ACACIA *decurrens dealbata*
ACACIA *decurrens mollis*
BAUHINIA *tomentosa*
CASSIA *fistula*
CASSIA *multijuga*
PARKINSONIA *aculeata*
THESPESIA *populnea*
TIPUANA *tipu*

CALLISTEMON *lanceolatus*
DELONIX *regia*
ERYTHRINA *var indica*
PLUMERIA *rubra*

APPENDIX A/1

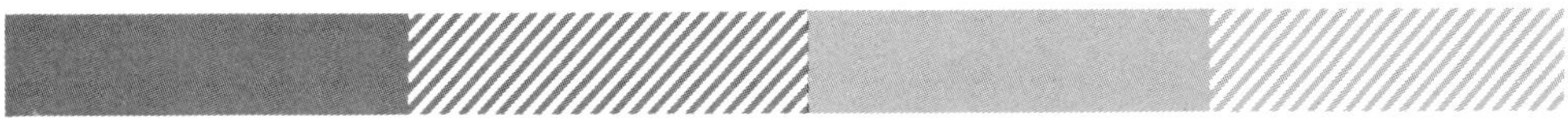

JACARANDA *mimosaefolia*	BAUHINIA *purpurea*	AZADIRACHTA *indica*	CASSIA *grandis*
	BAUHINIA *punctata*	BAUHINIA *alba var candida*	PONGAMIA
	GLIRICIDIA *maculata*	CORDIA *abyssinica*	SESBANIA *grandiflora*
	MELIA *azedarach*	MILLINGTONIA *hortensis*	TAMARIX *var.*
		MORINGA *pterygosperma*	
		PLUMERIA *acutifolia*	

Flowering & foliage shrubs suitable for the Middle East

**Key:-* see APPENDIX A

LATIN NAME	*KEY	PROPAGATION
ACACIA *farnesiana*	FH. FT.	Seed or cuttings
ACALYPHA *wilkesiana*	FH. T. O.	Cuttings
ARALIA *balfouri*	FH. T. O.	Cuttings
ATRIPLEX *halimus*	H. T.	Cuttings
BOUGAINVILLA *formosa*	FH. FT. O.	Cuttings
BUDDLEIA *alternifolia*	D.	Seed or cuttings
BUDDLEIA *asiatica*	FH. FT. O.	Seed or cuttings
BUDDLEIA *lindeyana*	D.	Cuttings
BUDDLEIA *variablis*	D.	Cuttings
BALERIA *prionitis*	FH. FT. O.	Cuttings
CAESALPINA *pulcherrima*	H. T. O.	Seed, cuttings or layers
CAESALPINA *gillesi*	H. T. O.	Seed or cuttings
CODIAEUM *var*	FH. FT. O.	Cuttings
DATURA *fastuosa*	FH. FT. O.	Seed or cuttings
DATURA *suaveolons*	FH. FT. O.	Seed or cuttings
DATURA *natalensis*	FH. FT. O.	Seed or cuttings
DODONAEA *viscosa*	H. FT.	Seed
DURANTA *repens*	FH. T. O.	Seed or cuttings
EUPHORBIA *pulcherrima*	D.	Cuttings
HIBISCUS *rosa-sinensis*	FH. FT. O.	Cuttings, grafts or marcots

APPENDIX B

REMARKS
Requires regular watering during establishment, light watering after, if necessary
Requires protection in early stages
Requires a semi-shaded position
Very hardy, tolerant
Requires protection from wind
Requires protection from wind
Requires some protection from wind
Requires protection from wind
Requires protection from wind
Requires protection in early stages
Should not be planted on sea front
Should not be planted on sea front
Requires some protection against wind
Seeds very poisonous
Seeds very poisonous
Seeds very poisonous
Requires regular watering during establishment, light watering after, if necessary
Requires some protection in early stages
Delicate, requires protection from wind
Requires some protection from wind

LATIN NAME	*KEY	PROPAGATION
HIBISCUS *schizopetalis*	FH. FT. O.	Cuttings
IXORA *coccinea*	FH. FT. O.	Cuttings
IXORA *chinensis*	FH. FT. O.	Cuttings
IXORA *javanica*	FH. FT. O.	Cuttings
LANTANA *var*	H. T. O.	Seed or cuttings
LAWSONIA *inermis*	H. T.	Seed or cuttings
MYRTUS *communis*	H. T. O.	Seed or cuttings
NERIUM *oleander*	H. T. O.	Seed or cuttings
OCIMUM *basilicum*	H. T.	Seed
PLUMBAGO *capensis*	D.	Seed or cuttings
PLUMBAGO *indica*	D.	Seed or cuttings
PLUMERIA *acutifolia*	FH. FT. O.	Cuttings
PLUMERIA *rubra*	FH. FT. O.	Cuttings
PUNICA *granatum*	H. T. O.	Seed, layering or grafting
RUSSELIA *juncea*	FH. FT.	Seed or cuttings
SESBANIA *aegyptica*	H. T.	Seed
TECOMA *stans*	H. T. O.	Seed or cuttings
TECOMA *smithii*	D.	Seed or cuttings
TECOMARIA *capensis*	D.	Seed or cuttings
YUCCA *filamentosa*	H. FT.	Suckers
YUCCA *gloriosa*	H. FT.	Suckers

REMARKS
Requires some protection from wind
Requires some protection from wind
Prefers a partially shaded situation
Requires some protection from wind
Requires regular maintenance
Flowers very highly scented
Should not be planted on sea front
Hardy and ornamental
Prefers semi-shaded situation
Delicate, requires protection from wind
Delicate, requires protection from Wind
Requires protection from wind
Requires protection from wind
Requires regular maintenance
Requires regular maintenance
Quick growing, annual/biennial
Quick growing ornamental
Slightly delicate, requires protection
Requires regular maintenance, fairly delicate
Requires regular watering during establishment, light watering after
Requires regular watering during establishment, light watering after

Classification of flowering & foliage shrubs according to colour

ARALIA *balfouri*	CAESALPINA *pulcherrima*	ACACIA *farnesiana*	ACALYPHA *wilkesiana*
ATRIPLEX *halimus*	IXORA *javanica*	BALERIA *prionitis*	EUPHORBIA *pulcherrima*
CODIAEM *var*	LANTANA *var*	CAESALPINA *pulcherrima*	HIBISCUS *rosasinensis*
DODONAEA *viscosa*	PUNICA *granatum*	CAESALPINA *gilliesii*	HIBISCUS *rosa-sinensis*
MYRTUS *communis*	TECOMA *smithii*	LANTANA *var*	IXORA *coccinea*
OCIMUM *basilicum*		NERIUM *oleander*	IXORA *chinens*
		SESBANIA *aegyptiaca*	LANTANA *var*
		TECOMA *smithii*	NERIUM *oleander*
		TECOMA *stans*	PLUMERIA *rubra*
			PUNICA *granatum*
			RUSSELIA *juncea*
			TECOMARIA *capensis*

APPENDIX B/1

DURANTA *repens*	BOUGAINVILLIA *formosa*	DATURA *fastuosa*	LANTANA *var*
PLUMBAGO *capensis*	BUDDLEIA *alternifolia*	DATURA *natalensis*	NERIUM *oleander*
	BUDDLEIA *asiatica*	DATURA *suaveolons*	PLUMBAGO *indica*
	BUDDLEIA *lindeyana*	LANTANA *var*	
	BUDDLEIA *variablis*	LAWSONIA *inermis*	
	DATURA *fastuosa*	NERIUM *oleander*	
		PLUMERIA *acutifolia*	
		YUCCA *filamentosa*	
		YUCCA *gloriosa*	

Climbers and Creepers suitable for the Middle East

**Key*:- see APPENDIX A

LATIN NAME	*KEY	PROPAGATION
ALLAMANDA *cathartica*	D. O.	Cuttings
ALLAMANDA *hendersonii*	D. O.	Cuttings or seed
ALLAMANDA *nerifolia*	D. O.	Cuttings
ANTIGONON *leptopus*	FH. FT. O.	Seed
ARISTOLOCHIA *elegans*	D. O.	Seed or cuttings
ASPARAGUS *plumosa*	FH. FT. O.	Seed or division
ASPARAGUS *sprengeri*	FH. FT. O.	Seed
BOUGAINVILLIA *var*	FH. FT. O.	Cuttings or plants
CLERODENDRON *indica*	HT. O.	Cuttings
CLITORIA *ternatea*	FH. O.	Seed
GLORIOSA *rothschildiana*	FH. O.	Seed or division
IPOMOEA *horsfalliae*	FH. O.	Seed or cuttings
IPOMOEA *learii*	H. T. O.	Seed or cuttings
IPOMOEA *pes caprea*	H. T.	Cuttings
IPOMOEA *purpurea*	H. FT. O.	Seed
IPOMOEA *quamoclit var pennata*	D. O.	Seed

APPENDIX C

REMARKS
Delicate, requires protection from wind
Delicate, requires protection from wind
Delicate, requires protection from wind
Requires protection in early stages
Requires protection from wind
Requires semi-shaded situation
Requires semi-shaded situation
Requires some protection from wind
Very hardy, excellent windbreak
Requires some protection from wind
Slow growing, requires some protection from wind
Slow growing, requires some protection from wind
Hardy, requires some protection from wind
Very hardy, saline tolerant
Requires some protection from wind
Requires protection from wind

LATIN NAME	*KEY	PROPAGATION
JACQUEMONTIA *martii*	FH. FT. O.	Seed, cuttings or layers
JASMINE *var*	FH. FT. O.	Seed, cuttings, or layers
QUISQUALIS *indica*	D. O.	Cuttings
SOLANUM *wendlandii*	D. O.	Cuttings
THUNBERGIA *alata*	FH. FT. O.	Seed
THUNBERGIA *grandiflora*	D. O.	Seed, cuttings, or tubers

REMARKS
Requires protection from wind
Requires protection from wind
Requires rich, well-drained soil and regular maintenance
Requires protection from wind
Usually grown as an annual
Requires protection from wind

Classification of Climbers & Creepers according to colour

ASPARAGUS *plumosa*	BOUGAINVILLIA *var*	ALLAMANDA *cathartica*	ANTIGONON *leptopus*
ASPARAGUS *sprengeri*		ALLAMANDA *hendersonii*	BOUGAINVILLIA *var*
CLERODENDRON *indica*		ALLAMANDA *nerifolia*	GLORIOSA *rothschildiana*
		BOUGAINVILLIA *var*	IPOMOEA *horsfalliae*
		JASMINE *revolutum*	IPOMOEA *quamoclit var pennata*
		JASMINE *primulinum*	QUISQUALIS *indica*
		THUNBERGIA *alata*	

APPENDIX C/1

CLITORIA *ternatea*	ARISTOLOCHIA *elegans*	BOUGAINVILLIA *var*	BOUGAINVILLIA *var*
JACQUEMONTIA *martii*	BOUGAINVILLIA *var*	JASMINE *officinale*	JASMINE *var*
	IPOMOEA *learii*	JASMINE *primulinum*	
	IPOMOEA *pes capre*	JASMINE *pubescens*	
	IPOMOEA *purpurea*	JASMINE *sambac*	
	SOLANUM *wendlandii*		
	THUNBERGIA *grandiflora*		

Tuberous & Herbaceous Plants suitable for the Middle East

**Key:-* see APPENDIX A

LATIN NAME	*KEY	PROPAGATION
AGAPANTHUS *africanus*	D. O.	Seed or division
AGAVE *americana*	H. T. O.	Offsets
AGAVE *sisalana*	H. T.	Offsets
ALOE *var*	H. FT. O.	Seed or division
CANNA *var*	H. T. O.	Seed or tubers
COLEUS *var*	FH. FT. O.	Seed
CRINUM *var*	FH. FT. O.	Division
DIEFENBACHIA *var*	FH. FT.	Stem-cuttings
EUCHARIS *amazonica*	FH. FT. O.	Bulbs
GLORIOSA *var*	FH. O.	Seed or division
HEMEROCALLIS *var*	FH. FT. O.	Division
HIPPEASTRUM *var*	FH. FT. O.	Seed and offshoots
HYLINE *gardneriana*	H. T. O.	Division or bulbs
HYMENOCALLIS *var*	FH. FT. O.	Offsets
KNIPHOFIA *var*	H. FT. O.	Division
MIRABLIS *julapa*	H. T. O.	Seed

APPENDIX D

REMARKS
Semi-shaded situation
Hardy, slow growing
Hardy, slow growing
Requires regular watering during establishment
Rich soil, high irrigation regime
Preferable in pots
Preferable in pots
Preferable in pots
Pots or beds
Slow growing, protection from wind
Beds or borders
Preferable in pots
Hardy, attractive, scented
Some varieties delicate
Regular irrigation regime
Spring planting

LATIN NAME	*KEY	PROPAGATION
OCIMUM *basilicum*	H. T.	Seed
RHOEO *discolour*	H. T. O.	Seed or division
SANSEVIERIA *trifasciata*	H. T.	Division
VINCA *rosea*	H. T. O.	Seed or cuttings
ZEBRINA *pendula*	H. T. O.	Cuttings

REMARKS
Hardy, tolerant
Prefers an open situation
Open situation
Open situation
Preferable in pots

Classification of Tuberous & Herbaceous plants according to colour

AGAVE *americana*	CANNA *lily*	GLORIOSA *var*	ALOE *var*
AGAVE *sisalana*	HEMEROCALLIS *var*	HEMEROCALLIS *var*	CANNA *lillies*
ALOE *var*	HIPPEASTRUM *var*	HIPPEASTRUM *var*	COLEUS *var*
DIEFENBACHIA *var*	KNIPHOFIA *var*	KNIPHOFIA *var*	GLORIOSA *var*
OCIMUM *basilicum*		MIRABILIS *jalapa*	KNIPHOFIA *var*
SANSIEVIERIA *trifasciata*			MIRABILIS *jalapa*

APPENDIX D/1

AGAPANTHUS	RHOEO	CRINUM	CANNA
africanus	*discolor*	*var*	*lillies*
	ZEBRINA	EUCHARIS	CRINUM
	pendula	*amazonica*	*var*
		HIPPEASTRUM	HIPPEASTRUM
		var	*var*
		HYLINE	MIRABLIS
		gardneriana	*jalapa*
		HYMENOCALLIS	VINCA
		var	*rosea*
		KNIPHOFIA	
		var	
		MIRABILIS	
		jalapa	

Windbreaks, Screening & Boundaries suitable for the Middle East

**Key*:- see APPENDIX A

LATIN NAME	*KEY	PROPAGATION
ACACIA *arabica*	H. T. O.	Seed
ACACIA *decurrens mollis*	H. T. O.	Seed or cuttings
ACACIA *farnesiana*	H. T. O.	Seed or cuttings
AGAVE *americana*	H. T. O.	Offsets
AGAVE *sisalana*	H. T.	Offsets
CASUARINA *cunninghamiana*	H. T. O.	Seed
CASUARINA *equisetifolia*	H. T.	Seed
CLERODENDRON *indica*	H. T. O.	Cuttings
DODONEA *viscosa*	H. T. O.	Seed
FIMBRISTYLIS *var*	FH. T.	Separation
GLIRICIDIA *maculata*	FH. FT. O.	Seed or cuttings
LEUCAENA *glauca*	FH. FT.	Seed
MELIA *azedarach*	H. T. O.	Seed or cuttings
NERIUM *oleander*	H. T. O.	Seed or cuttings
OPUNTIA *var*	H. T.	Cuttings
PARKINSONIA *aculcata*	FH. T. O.	Seed

APPENDIX E

REMARKS
Suitable desert conditions, screening and windbreaks
Suitable desert conditions, screening and surround planting
Suitable desert conditions, screening
Suitable desert conditions, boundary planting
Suitable desert conditions, boundary planting
Suitable desert conditions, screening, windbreaks
Suitable desert conditions, screening, windbreaks and coastal protection
Suitable desert conditions, screening and windbreaks
Suitable desert conditions, screening and boundary planting
Suitable for windbreaks and screening when planted on irrigation channels
Suitable for screening, especially if planted on irrigation channels
Suitable desert conditions, screening and windbreaks
Suitable for screening in protected environment
Suitable for screening, regular watering with high requirement
Suitable desert conditions, boundary planting
Suitable for screening in partially protected environment

LATIN NAME	*KEY	PROPAGATION
PHRAGMITES *communis*	H. T.	Division
PITHECOLOBIUM *dulce*	H. T. O.	Seed
POPULUS *alba*	H. FT. O.	Seed or cuttings
PROSOPIS *juliflora*	H. T.	Seed
SCHINUS *molle*	H. T. O.	Seed or cuttings
SCHINUS *terebinthifolius*	FH. FT. O.	Seed or cuttings
SESBANIA *aegyptiaca*	H. T.	Seed
SESBANIA *grandiflora*	FH. T. O.	Seed
TAMARIX *var*	H. T.	Cuttings
TECOMA *stans*	FH. FT. O.	Seed or cuttings
TERMINALIA *catappa*	H. T. O.	Seed
ZIZYPHUS *spina-christi*	H. T.	Seed or cuttings
ZIZYPHUS *vulgare*	H. T.	Seed

REMARKS
Suitable for screening or windbreaks when planted on water channels
Suitable for screening, windbreaks or closely planted as impenetrable hedge
Suitable for screening and windbreaks in suitable environment
Suitable for desert conditions, screening, windbreaks and boundary planting
Suitable for screening in suitable environment
Suitable for screening in partially protected environment
Suitable for screening or light windbreaks
Suitable for screening especially in coastal areas
Suitable desert and coastal conditions, screening and windbreaks
Suitable for screening in partially protected environment
Suitable under light irrigation for screening
Suitable desert conditions, screening, windbreaks and boundary planting
Suitable desert conditions, screening and windbreaks

Ornamental hedges suitable for the Middle East

**Key:-* see APPENDIX A

LATIN NAME	*KEY	PROPAGATION
ACACIA *farnesiana*	H. T. O.	Seed or cuttings
ACALYPHA *wilkesiana*	FH. FT. O.	Cuttings
ACALYPHA *var*	FH. FT. O.	Cuttings
AGAVE *americana*	H. T. O.	Offsets
ARALIA *balfouri*	FH. FT. O.	Cuttings
ATRIPLEX *halimus*	H. T.	Cuttings
BARLERIA *prionitis*	FH. FT. O.	Cuttings
BOUGAINVILLIA *var*	FH. FT. O.	Cuttings
CAESALPINA *pulcherrima*	FH. FT. O.	Seed, cuttings or layers
CLERONDENDRON *indica*	H. T. O.	Cuttings
CODIAEUM *var*	FH. FT. O.	Cuttings
DODONEA *viscosa*	H. T. O.	Seed
DURANTA *repens*	FH. T. O.	Seed or cuttings
HIBISCUS *rosa-sinensis*	FH. FT. O.	Cuttings, grafts or marcots
HIBISCUS *schizopetalus*	FH. FT. O.	Cuttings, grafts or marcots
LANTANA *var*	H. T. O.	Seed or cuttings

APPENDIX F

REMARKS

Suitable desert conditions, hedges

Ornamental hedges, responds well to pruning

Ornamental hedges, responds well to pruning

Ornamental hedge when planted in depth

Ornamental hedge in shaded and protected situation

Suitable desert conditions, low hedges

Ornamental hedge in partially protected situation

Ornamental hedge, sunny, protected situation

Ornamental hedge, rapid growth in suitable environment

Ornamental hedge, rapid growth rate, eminently suitable in coastal areas

Ornamental hedge in semi-shaded and protected situation

Suitable desert conditions, screening and boundary hedges

Ornamental hedge, sunny, partially protected position

Ornamental hedge, sunny, protected situation

Ornamental hedge, sunny protected situation

Ornamental hedge requiring regular maintenance

LATIN NAME	*KEY	PROPAGATION
LAWSONIA *inermis*	FH. T. O.	Seed or cuttings
MYRTUS *communis*	FH. T. O.	Seed or cuttings
NERIUM *oleander*	H. T. O.	Seed or cuttings
PARKINSONIA *aculeata*	FH. T. O.	Seed
PITHECOLOBIUM *dulce*	H. T. O.	Seed
PUNICA *granatum*	H. T. O.	Seed, layering or grafting
TECOMA *stans*	FH. FT. O.	Seed or cuttings
TECOMA *smithii*	D. FT. O.	Seed or cuttings

REMARKS

Froms a straggly, highly scented hedge in partially protected areas

Forms an attractive, fragrant hedge in partially protected position

Ornamental hedge, regular watering with high requirement

Ornamental hedge, with regular maintenance, partially protected position

Impenetrable hedge, if closely planted, requires regular cutting back

Ornamental hedge, tends to become straggly, requires regular maintenance

Forms an attractive hedge in a partially protected environment

Forms an attractive hedge in a sunny, protected situation

Annuals suitable for planting in the Middle East
Salt tolerance table

APPENDIX G

FAIRLY HIGH TOLERANCE	MODERATE TOLERANCE	POOR TOLERANCE
AGERATUM	ANTIRRHINUM	AQUILEGIA
ALYSUM	ARCTOTIS	BRACHYCOME
AMARANTHUS	BLUE LACE FLOWER	CLIANTHUS
ASPARAGUS PLUMOSA	CANARY CREEPER	PHLOX DRUMMONDI
ASPARAGUS SPRENGERI	CANDYTUFT	SALPIGLOSSIS
BALSAM	CLEOME	SWEET WILLIAM
CACTUS	COSMOS	SWEET PEA
CALENDULA	SALVIA	
CALLIOPSIS	GLORIOSA DAISY	
CANNA		
CAPSICUM ORNAMENTAL		
CARNATION		
CELOSIA		
CHRYSANTHEMUM		
COLEUS		
CONVOLVULUS		
CORNFLOWER		
DAHLIA		
DIANTHUS		
FOUR-O-CLOCK		
GAILLARDIA		
GLOBE AMARANTH		
HELICHRYSUM		
HOLLYHOCK		
KOCHIA		
LARKSPUR		
LINUM		
MARIGOLD VAR		
NASTURTIUM		
NICOTIANA		
PETUNIA		
STATICE		

FAIRLY HIGH TOLERANCE	MODERATE TOLERANCE	POOR TOLERANCE
STOCK		
SUNFLOWER		
TITHONIA		
VERBENA		
ZINNIA		
MESEMBRYANTHEMUM		

Provisional planting guide – Middle East

APPENDIX H

SEEDS THAT CAN BE PLANTED THROUGHOUT THE YEAR	SEEDS THAT CAN BE PLANTED IN SPRING	SEEDS THAT CAN BE PLANTED IN AUTUMN
CANNAS	ASPARAGUS PLUMOSA	AGERATUM
PERIWINKLE	ASPARAGUS SPRENGERI	ALYSSUM
ZINNIA	CAPSICUM ORNAMENTAL	AMARANTHUS
	CELOSIA	ANTIRRHINUM
	CLEOME	AQUILEGIA
	COLEUS	ARCTOTIS
	FOUR O'CLOCK	ASPARAGUS PLUMOSA
	GAILLARDIA	ASPARAGUS SPRENGERI
	GLOBE AMARANTH	BALSAM
	GLORIOSA DAISY	BLUE LACE FLOWER
	KOCHIA	CALENDULA
	MIMOSA PUDICA	CALLIOPSIS
	PETUNIA	CANDYTUFT
		CARNATION
		CELOSIA
		CHRYSANTHEMUM
		CLEOME
		CLIANTHUS
		COLEUS
		CONVOLVULUS
		CORNFLOWER
		COSMOS
		DAHLIA
		DIANTHUS
		GAILLARDIA
		HELICHRYSUM
		HOLLYHOCK
		IPOMOEA
		LARKSPUR
		LINARIA
		LINUM

SEEDS THAT CAN BE PLANTED THROUGHOUT THE YEAR	SEEDS THAT CAN BE PLANTED IN SPRING	SEEDS THAT CAN BE PLANTED IN AUTUMN
		MARIGOLD
		NASTURTIUM
		NICOTIANA
		PETUNIA
		PHLOX DRUMMONDI
		SALVIA
		STATICE
		STOCK
		SUNFLOWER
		SWEET SULTAN
		SWEET PEA
		THUNBERGIA
		TITHONIA
		VERBENA

Index of plants

LATIN NAME	FAMILY	Page No.
SCORZONERA *judaica*	Compositae	99
SESBANIA *aegyptiaca*	Papilionaceae	49
grandiflora		27
SOLANUM *wendlandii*	Solanaceae	61
SPATHODEA *campanulata*	Bignoniaceae	29
nilocta		29
STIPA *tortilis*	Graminae	89
TAMARINDUS *indica*	Caesalpiniaceae	29
TAMARIX *var*	Tamaricaceae	29
TECOMA *smithii*	Bignoniaceae	49
stans		49
TECOMARIA *capensis*		50
TERMINALIA *catappa*	Combretaceae	31
TETRADICLIS *salsa*	Zygophyllaceae	100
THESPESIA *populnea*	Malvaceae	31
THEVETIA *nerefolia*	Apocynaceae	31
THUNBERGIA *alata*	Acanthaceae	63
grandiflora		63
TIPUANA *tipu*	Papilionaceae	31
URTICA *urens*	Urticaceae	81
VINCA *rosea*	Apocynaceae	75
YUCCA *filamentosa*	Liliaceae	50
gloriosa		50
ZEBRINA *pendula*	Commelinaceae	75
ZIZYPHUS *spina-christi*	Rhamnaceae	32
vulgare		32

Index of plants

COMMON NAME	LATIN NAME	Page No.
Bowstring hemp	SANSEVIERIA *trifasciata*	75
Buffalo grass	CENCHURUS *ciliaris*	86
Burstwort	HERNIARIA *hemistemon*	79
Butterfly bush	BUDDLEIA *alternifolia*	36
	asiatica	36
	lindeyana	36
	variablis	37
Cactus	CACTUS *var*	94
Calico flower	ARISTOLOCHIA *elegans*	54
Camel's foot	BAUHINIA *alba var candida*	5
	punctata	7
	purpurea	5
	tomentosa	7
Canna lily	CANNA *var*	69
Cape honeysuckle	TECOMARIA *capensis*	50
Carob bean	CERATONIA *siliqua*	11
Carrichtera	CARRICHTERA *annual*	95
Cassia-horse	CASSIA *grandis*	9
-leafy	*multijuga*	9
Cassie	ACACIA *farnesiana*	35
Century plant	AGAVE *americana*	67, 93
Cistanche	CISTANCHE *lutea*	95
Clematis-red	IPOMOEA *horsfailliae*	57
Clerodendron	CLERODENDRON *indica*	55
Clitoria	CLITORIA *ternatea*	55
Coleus	COLEUS *var*	69
Common reed	PHRAGMITES *communis*	99
Coral creeper	ANTIGONON *leptopus*	53
hibiscus	HIBISCUS *schizopetalus*	41
plant	RUSSELIA *juncea*	49
tree	ERYTHRINA *indica*	13
Cork tree-Indian	MILLINGTONIA *hortensis*	21
Country almond	TERMINALIA *catappa*	31
Crinum lily	CRINUM *var*	69
Croton	CODIAEUM *var*	39
Crown-of-thorns	ZIZYPHUS *spina-christi*	32
Cymbopogon	CYMBOPOGON *schoenanthus*	86
Cypress vine	IPOMOEA *quamoclit var pennata*	59
Dactyloctenium	DACTYLOCTENIUM *aegyptiacum*	87
Date palm	PHOENIX *dactylifera*	21
Datura	DATURA *fastuosa*	39
	natalensis	40

General index